SpringerBriefs in Applied Sciences and Technology

SpringerBriefs present concise summaries of cutting-edge research and practical applications across a wide spectrum of fields. Featuring compact volumes of 50–125 pages, the series covers a range of content from professional to academic.

Typical publications can be:

- A timely report of state-of-the art methods
- An introduction to or a manual for the application of mathematical or computer techniques
- A bridge between new research results, as published in journal articles
- A snapshot of a hot or emerging topic
- An in-depth case study
- A presentation of core concepts that students must understand in order to make independent contributions

SpringerBriefs are characterized by fast, global electronic dissemination, standard publishing contracts, standardized manuscript preparation and formatting guidelines, and expedited production schedules.

On the one hand, **SpringerBriefs in Applied Sciences and Technology** are devoted to the publication of fundamentals and applications within the different classical engineering disciplines as well as in interdisciplinary fields that recently emerged between these areas. On the other hand, as the boundary separating fundamental research and applied technology is more and more dissolving, this series is particularly open to trans-disciplinary topics between fundamental science and engineering.

Indexed by EI-Compendex, SCOPUS and Springerlink.

More information about this series at http://www.springer.com/series/8884

Afshin J. Ghajar

Two-Phase Gas-Liquid Flow in Pipes with Different Orientations

 Springer

Afshin J. Ghajar
School of Mechanical and Aerospace Engineering
Oklahoma State University
Stillwater, OK, USA

ISSN 2191-530X   ISSN 2191-5318   (electronic)
SpringerBriefs in Applied Sciences and Technology
ISBN 978-3-030-41625-6   ISBN 978-3-030-41626-3   (eBook)
https://doi.org/10.1007/978-3-030-41626-3

This Springer imprint is published by the registered company Springer Nature Switzerland AG
The registered company address is: Gewerbestrasse 11, 6330 Cham, Switzerland

Preface

The phenomenon of gas–liquid two-phase flow in inclined systems, although not as common as horizontal or vertical flow, is of great practical significance in several applications such as undulating oil-gas flow lines, chemical process engineering, and inclined flow paths in steam condensers and generators. In these practical applications, accurate determination of two-phase flow variables such as void fraction, pressure drop, and heat transfer is of great importance for system sizing and optimization. It is a well-established fact that these parameters are very sensitive to the spatial and morphological variations of the two-phase flow structure. The two-phase flow structure, commonly termed as flow pattern, depends on the interaction and balance between the buoyancy-inertia-gravity forces which in turn are a function of pipe inclination. Thus, a correct understanding of the effect of change in pipe inclination on the two-phase flow structure is needed, and its effect on the thermofluidics of two-phase flow needs to be understood. Thus, the scope of this Springer Brief is to present an overview of the pipe inclination effects of the gas–liquid two-phase flow phenomenon.

The pipe inclination effects on non-boiling two-phase flow are studied through the data available in literature and extensive experiments carried out at the Two-Phase Flow Laboratory, Oklahoma State University, Stillwater, Oklahoma. These experiments were carried out using air-water as fluid combination. The experimental setup used for the experiments is unique as it can be rotated to any inclination between +90° and −90° including the horizontal orientation and is capable of flow visualization and simultaneous measurements of void fraction, pressure drop, and heat transfer. The experimental data show the flow pattern and pipe inclination dependency of all two-phase flow variables. At lower mass fluxes, this effect is found to be most significant where the buoyant forces acting on the gas phase dominate the two-phase flow phenomenon. Experiments also reveal the prevalence and insights about the flow reversal and transient nature of the two-phase flow in upward and downward inclined systems.

Two-phase flow literature reports a plethora of correlations/models for determination of void fraction, pressure drop, and non-boiling heat transfer. Since the two-phase flow is a function of several variables such as flow patterns (phase flow rates), fluid properties, pipe diameter, and inclination angle (pipe orientation), it is quite a challenging task for the end user to select an appropriate flow condition-specific correlation/model. Selection of a correct model also requires some fundamental understanding of the two-phase flow physics and the underlying principles/ assumptions/limitations associated with these correlations. To address these issues, this Springer Brief also introduces some of the top performing two-phase flow models validated against a comprehensive data set.

Stillwater, OK, USA   Afshin J. Ghajar

Acknowledgments

I would like to thank several of my colleagues around the world for sharing their two-phase flow experimental data with me. Majority of the results presented in this article is based on the work of my former students Dr. Swanand M. Bhagwat of Southwest Research Institute, San Antonio, Texas, and Professor Clement C. Tang of the University of North Dakota, Grand Forks. Their contributions are greatly appreciated.

Contents

Nomenclature

a_1	Variable in Xiong and Chung (2006) correlation
A	Cross-sectional area (m^2)
B_1	Variable defined by Eq. (5.28)
B_2	Variable defined by Eq. (5.29)
B_3	Variable defined by Eq. (5.30)
Bo	Bond number defined by Eq. (5.35)
c_1	Variable in Bhagwat and Ghajar (2014) correlation
c_2	Variable in Bhagwat and Ghajar (2014) correlation
c_3	Variable in Bhagwat and Ghajar (2014) correlation
C	Constant in Lockhart and Martinelli (1949) parameter
C_o	Distribution parameter
$C_{o,1}$	Variable used by Bhagwat and Ghajar (2014)
C_1	Variable in Bhagwat and Ghajar (2015a) correlation
C_2	Variable in Bhagwat and Ghajar (2015a) correlation
C_3	Variable in Bhagwat and Ghajar (2015a) correlation
C_4	Variable in Bhagwat and Ghajar (2015a) correlation
d_{def}	Bubble diameter above which bubble is deformed (m)
d_{max}	Maximum bubble diameter (m)
d_{migr}	Bubble diameter below which migration of bubbles is prevented (m)
dp/dz	Pressure gradient (Pa/m)
$(dp/dz)^+$	Non-dimensional pressure gradient
D	Inside pipe diameter (m)
D_h	Hydraulic diameter (m)
D^+	Non-dimensional pipe diameter used by Bhagwat and Ghajar (2015a)
D_h^+	Non-dimensional hydraulic diameter used by Kataoka and Ishii (1987)
E	Liquid entrainment fraction
Eo	Eötvös number
f	Fanning friction factor
F	Froude number defined by Eq. (3.10)

F_p	Flow pattern factor
F_s	Shape factor
Fr	Modified Froude number defined by Eq. (3.16)
g	Acceleration due to gravity (m/s^2)
G	Mass flux (kg/m^2 - s)
h_L	Liquid level in stratified flow (m), single-phase heat transfer coefficient (W/m^2 - K)
h_{TP}	Two-phase heat transfer coefficient (W/m^2 - K)
$\bar{h}$	Average heat transfer coefficient (W/m^2 - K)
I	Inclination factor
k	Thermal conductivity (W/m-K)
L	Test section length (m)
La	Laplace number
$\dot{m}$	Mass flow rate (kg/s)
MARD	Mean absolute relative deviation
MRD	Mean relative deviation
N	Number of data points
N_{30}	Percentage of data points predicted within $\pm 30\%$
N_{50}	Percentage of data points predicted within $\pm 50\%$
N_μ	Viscosity number
N_{ST}	Number of thermocouple stations
p	Pressure (Pa)
Pr	Prandtl number
$\dot{q}$	Heat transfer rate (W)
Re	Reynolds number
S	Slip ratio
T_b	Bulk temperature ($^\circ$C)
$\bar{T}_{w,i}$	Average inside wall temperature ($^\circ$C)
$\bar{T}_{w,o}$	Average outside wall temperature ($^\circ$C)
U	Phase velocity (m/s)
U_b	Bubble velocity (m/s)
U_{GM}	Drift velocity (m/s)
U_{GM}^+	Non-dimensional drift velocity used by Kataoka and Ishii (1987)
$U_{SG,t}$	Superficial gas velocity at smooth to wavy stratified flow transition (m/s)
V	Volume (m^3)
We	Weber number
x	Two-phase flow quality
X	Lockhart and Martinelli (1949) parameter
Y	Taitel and Dukler (1976) parameter, Chisholm (1973) parameter
Δz	Differential pipe length (between two thermocouple stations)

Greek Symbols

α_G	Void fraction
α_L	Liquid holdup
β	Volume fraction of liquid film
δ	Liquid film thickness in annular flow (m)
$\Delta\rho$	Density difference (kg/m^3)
ε	Roughness (m)
μ	Dynamic viscosity (Pa-s)
ν	Kinematic viscosity (m^2/s)
σ	Surface tension (N/m)
ρ	Density (kg/m^3)
λ	Gas volumetric flow fraction
θ	Pipe orientation (degree)
Φ^2	Two-phase frictional multiplier
ξ	Correction factor in Eq. (6.6)
π_1	Non-dimensional variable defined by Eq. (5.31)
π_2	Constant defined by Eq. (5.32)
π_3	Non-dimensional variable defined by Eq. (5.33)

Subscripts

a	Accelerational
atm	Atmospheric
c	Gas core
C	Critical
CAL	Calculated
EXP	Experimental
f	Frictional
G	Gas
GO	Gas only
h	Hydrostatic
i	Interface or inside
j	Phase
L	Liquid
LF	Liquid film
LO	Liquid only
max	Maximum
M	Mixture
o	Outside

ref	Reference
S	Superficial
sys	System
t	Total
tt	Turbulent-turbulent
TP	Two-phase
w	Wall

Superscripts

| $\sim$ | Non-dimensional quantities in Taitel and Dukler (1976) model |
| + | Non-dimensional quantities in Kataoka and Ishii (1987) model |

Chapter 1
Introduction

Background

The term two-phase flow designates a flow situation where two distinct phases flow simultaneously usually through a pipe. The two phases can be a combination of any two phases (gas-liquid, liquid-solid, gas-solid, or even two immiscible liquids). However, the main focus of the research work presented here is the simultaneous flow of gas-liquid two-phase flow in round pipes. The two-phase flow of gas and liquid is classified as one-component and two-component two-phase flow. One-component two-phase flow consists of the two phases of a single chemical species such as steam-water and refrigerant vapor-liquid flow used in nuclear and refrigeration applications, whereas, two-component two-phase flow consists of the two phases of two chemically different species such as air-water and air-oil. The two-component two-phase flow is often referred to as non-boiling two-phase flow and is usually encountered in industrial applications such as artificial lift systems and simultaneous transportation of oil and natural gas from remote extraction locations to the processing units. Chemical operations requiring flow of two chemical species together for enhanced mass transfer, as in case of ozone treatment of water, also rely on the two-component two-phase flow phenomenon. For the purpose of flow assurance and designing, sizing, and optimization of industrial processes and equipment handling two-phase flow, determination of the void fraction, total two-phase pressure drop, and heat transfer is crucial. In particular, for non-boiling two-phase flow applications, the total two-phase pressure drop based on flow patterns, void fraction, and pipe geometry influences the design of a two-phase flow system. Knowledge of non-boiling heat transfer in two-component two-phase flows is equally important to estimate and prevent advent of wax deposition in subsea pipeline carrying hydrocarbons.

Depending on the mass flow rates of individual phases, pipe orientation, and pipe diameter, the two phases align themselves in a particular fashion across the pipe cross section identified as a flow pattern. The principal problem associated with the gas-liquid two-phase flow is its flow pattern-dependent complex structure due to turbulent mixing, compressibility nature of the gas phase, significantly different thermo-physical properties of the two phases, phase flow rates, and the pipe orientation that renders the analytical relationships for void fraction, two-phase pressure drop, and heat transfer difficult. Thus, it is necessary to study the gas-liquid two-phase flow phenomenon experimentally and device a systematic method to correctly predict the two-phase flow parameters such as void fraction, pressure drop, and heat transfer.

Two-phase flow literature reports a plethora of correlations for determination of flow patterns, void fraction, two-phase pressure drop, and non-boiling heat transfer correlations; however, the validity of a majority of these correlations is restricted over a narrow range of two-phase flow conditions. Consequently, it is quite a challenging task for the end user to select an appropriate correlation/model for the type of two-phase flow under consideration. Selection of a correct correlation also requires some fundamental understanding of the two-phase flow physics and the underlying principles/assumptions/limitations associated with these correlations. Thus, it is of significant interest for a design engineer to have knowledge of the flow patterns and their transitions and their influence on two-phase flow variables.

Need of Two-Phase Flow Study in Inclined Systems

The research in the field of gas-liquid two-phase flow can be classified in terms of its relevance to flow patterns, void fraction, two-phase pressure drop, and heat transfer. The accurate prediction of void fraction is very important in correct estimation of two-phase mixture density and hence the two-phase hydrostatic pressure drop and the convective heat transfer coefficient. The two-phase pressure drop essentially varies from that of its single-phase counterpart due to the significant density difference and hence the slippage between the two phases. Although, the two-phase flow literature reports a number of void fraction and pressure drop correlations, however, they are found to be inadequate in their range of application in terms of flow patterns, pipe orientation, and the fluid physical properties. Most of these correlations are of empirical form and are developed based on limited experimental data and hence perform well only over the certain range of two-phase flow conditions.

In comparison to the horizontal and vertical pipe orientations, the two-phase flow literature is devoid of flow patterns, void fraction, two-phase pressure drop, and heat transfer data in upward and downward inclined pipes. Consequently, the accuracy and performance of the existing two-phase flow correlations for the case of upward and downward inclined two-phase flow is unknown and uncertain. Moreover, due to lack of experimental initiatives, the two-phase flow literature lacks sound understanding of the effect of pipe orientation on two-phase flow parameters. This uncertainty and lack of sound understanding of the inclined two-phase flow phenomenon possess a challenge in the selection procedure of the most appropriate correlation to be used for upward or downward inclined two-phase flow situations.

Thus, this creates a room for further experimental investigation of two-phase flow parameters for upward and downward inclined pipe orientations and contribution to our fundamental understanding of the two-phase flow mechanism. Additionally, due to wide industrial applications of two-phase flow and high uncertainty in correct prediction of flow patterns, it is strongly desired to have unified two-phase flow models that can predict two-phase flow parameters such as void fraction, two-phase pressure drop, and heat transfer independent of flow patterns, pipe orientation, and thermo-physical properties of gas and liquid phases.

Basic Definitions in Gas Liquid Two-Phase Flow

The gas-liquid two-phase flow is realistically a three-dimensional flow with flow conditions and fluid properties varying with respect to the pipe cross section, length, and time. Moreover, the simultaneous gas-liquid two-phase flow is marked by the continuous interaction between the two phases with a dynamic interface along the pipe length and cross section and thus making it difficult to model two-phase flow phenomenon mathematically. Thus, this complexity limits the application of conventional equations designed for single-phase flow to the two-phase flow situations. In order to have an understanding of the two-phase flow parameters and its deviation from single-phase flow, basic definitions of two-phase flow are presented in this chapter. A valid assumption of one-dimensional two-phase flow to a great extent simplifies several two-phase flow variables and modeling techniques.

Table 1.1 introduces some of the primary variables that have been adopted by different two-phase flow models. The parameters listed in Table 1.1 are defined in context to single-phase flow of gas and liquid phase and their simple relationships that represent some two-phase flow variables. The mass flow rate of the gas ($\dot{m}_G$) and liquid ($\dot{m}_L$) phase through the pipe can be measured directly by mass flow meters. Based on these mass flow rates, the total or two-phase mixture mass flow rate ($\dot{m}$) is defined as the sum of the mass flow rates of individual phases. The two-phase flow quality (x) is defined as the ratio of mass flow rate of the gas phase ($\dot{m}_G$) to the two-phase mixture mass flow rate ($\dot{m}$). One of the important two-phase flow variables is the void fraction (α_G) and can be defined in different ways based upon its measurement technique such as cross-sectional or volumetric method. For example, if the void fraction is measured using cross-sectional (conductance/capacitance probe) technique, then it is defined as the ratio of pipe cross-sectional area occupied by the gas phase (A_G) to the total pipe cross-sectional area ($A = A_G + A_L$). Whereas in the case of volumetric void fraction measurement using quick closing valves, the void fraction is defined as the ratio of volume of the pipe occupied by the gas phase (V_G) to the total volume of test section (V). Another important two-phase flow parameter that is usually used in the analysis of two-phase flow is the gas volumetric flow fraction (λ) which is essentially the homogeneous void fraction under the

Table 1.1 Basic definitions of interest in two-phase flow

Parameter	Definition	Description
$\dot{m}$	$\dot{m} = \dot{m}_G + \dot{m}_L$	Mixture mass flow rate (kg/s)
x	$\frac{\dot{m}_G}{\dot{m}_G + \dot{m}_L}$	Two-phase flow quality
α_G	$\frac{A_G}{A}$ or $1 - \frac{A_L}{A}$, $\frac{V_G}{V}$ or $1 - \frac{V_L}{V}$	Void fraction
λ	$\frac{U_{SG}}{U_{SG} + U_{SL}}$ or $\left[1 + \frac{1-x}{x}\left(\frac{\rho_G}{\rho_L}\right)\right]^{-1}$	Gas volumetric flow fraction
U_{SG}	$\frac{Gx}{\rho_G}$ or $\frac{\dot{m}_G}{A\rho_G}$	Superficial gas velocity (m/s)
U_{SL}	$\frac{G(1-x)}{\rho_L}$ or $\frac{\dot{m}_L}{A\rho_L}$	Superficial liquid velocity (m/s)
U_G	$\frac{U_{SG}}{\alpha_G}$ or $\frac{\dot{m}_G}{A_G\rho_G}$	Actual gas velocity (m/s)
U_L	$\frac{U_{SL}}{1-\alpha_G}$ or $\frac{\dot{m}_L}{A_L\rho_L}$	Actual liquid velocity (m/s)
U_M	$U_{SL} + U_{SG}$ or $U_L(1 - \alpha_G) + U_G\alpha_G$	Two-phase mixture velocity (m/s)
S	$\frac{U_G}{U_L}$ or $\left(\frac{x}{1-x}\right)\left(\frac{1-\alpha_G}{\alpha_G}\right)\left(\frac{\rho_L}{\rho_G}\right)$	Slip ratio
Re_{SL}	$\frac{\rho_L U_{SL} D}{\mu_L}$ or $\frac{G(1-x)D}{\mu_L}$	Superficial liquid Reynolds number
Re_{SG}	$\frac{\rho_G U_{SG} D}{\mu_G}$ or $\frac{GxD}{\mu_G}$	Superficial gas Reynolds number

assumption of no slippage between the two phases. It is defined as the ratio of the superficial gas velocity (U_{SG}) to the mixture velocity ($U_{SG} + U_{SL}$). Velocity of liquid and gas phase assuming that phase is flowing alone through the pipe occupying the entire cross section is called as phase superficial velocity. The superficial gas or liquid velocity (U_{SG} or U_{SL}) is defined assuming the phase mass flow rate/mass flux (G) through the entire pipe cross section (A). Actual phase velocity (U_G or U_L) is quite higher than that of the superficial phase velocity due to the reduced pipe cross-sectional area available for each phase. The actual phase velocity is based on the phase mass flow rate through the pipe cross-sectional area (A_G or A_L) occupied by that phase. The two-phase mixture velocity (U_M) is simply the summation of the superficial velocity of each phase. The slip ratio (S) is the ratio of actual gas phase velocity (U_G) to actual liquid phase velocity (U_L). Finally, the superficial liquid or gas Reynolds number (Re_{SL} or Re_{SG}) is based on the phase superficial velocity (U_{SG} or U_{SL}), phase density (ρ_L or ρ_G), and phase dynamic viscosity (μ_L or μ_G).

Chapter 2
Two-Phase Flow Experimental Setup for Inclined Systems

Background

As mentioned earlier in Chap. 1, two-phase flow literature reports a plethora of correlations for determination of flow patterns, void fraction, two-phase pressure drop, and non-boiling heat transfer correlations. However, majority of these correlations were developed based on experimental data that covered a limited range of pipe orientations, e.g., vertical upward and horizontal, and certain flow patterns, e.g., the annular flow regime. In order to understand the phenomenon and physics of void fraction, pressure drop, and non-boiling heat transfer in two- phase flow and study its dependency on various flow parameters, it is necessary to carry out systematic experimental measurements. The experimental work presented in this study intends to complement the two-phase flow data available in the literature and aid to analyze the behavior of void fraction, two-phase pressure drop, and non-boiling heat transfer with varying flow patterns and pipe orientations. Once the necessary experimental data has been collected and the nexus and interdependencies between flow patterns, void fraction, two-phase pressure drop, and non-boiling heat transfer have been established, the next major move to accurately quantify the void fraction, two-phase pressure drop, and non-boiling heat transfer is the development of the two-phase flow models.

In order to have a sound understanding of the effect of pipe orientation on two-phase flow phenomenon, the collected experimental data should help in addressing the following issues:

1. What is the qualitative and quantitative effect of pipe orientation on the physical structure of flow patterns and their transition boundaries?
2. What is the relationship between flow patterns and void fraction as a function of pipe orientation?
3. How sensitive is the two-phase pressure drop to the void fraction for different flow patterns at any given pipe orientation?

A. J. Ghajar, *Two-Phase Gas-Liquid Flow in Pipes with Different Orientations*, SpringerBriefs in Applied Sciences and Technology, https://doi.org/10.1007/978-3-030-41626-3_2

4. What is the effect of flow pattern structure and their transition from one to the other on the frictional component of the two-phase pressure drop?
5. What is the effect of pipe surface roughness on the frictional component of two-phase pressure drop for different flow patterns at any given pipe orientation?
6. What is the effect of change in gas and liquid flow rates in different pipe inclinations on the two-phase heat transfer coefficient?
7. What is the effect of pipe inclination on the two-phase heat transfer coefficient?
8. What is the most reliable and accurate method to predict void fraction, pressure drop, and heat transfer in upward and downward inclined gas–liquid two-phase flow?

Details of the Two-Phase Flow Experimental Setup

The versatile experimental setup described here is located at the School of Mechanical and Aerospace Engineering, Two-Phase Flow Laboratory at Oklahoma State University (OSU), Stillwater, Oklahoma. The setup can be used for flow visualization and measurements of void fraction, pressure drop, and non-boiling heat transfer for different pipe orientations (+90° to −90°), all using a single experimental setup.

The experimental setup used in the present study consists of two test sections: a 12.7 mm I.D. polycarbonate transparent pipe section and a 12.5 mm I.D. stainless steel pipe. The experimental setup is mounted on a variable inclination frame (made up of rectangular steel tubing) with the help of pulleys and bolts and thus making it flexible to be rotated to any inclination between +90° and −90° including the horizontal orientation. The test section of 12.7 mm I.D. and 890-mm-long transparent polycarbonate pipe is used for flow visualization, void fraction, and pressure drop measurements, while 12.5 mm I.D. and 1030-mm-long schedule 40S stainless steel pipe having a surface roughness of 0.02 mm is used for pressure drop and heat transfer measurements. An overview of the experimental setup and the variable inclination frame is shown in Figs. 2.1 and 2.2, respectively.

The fluid combination used for generating two-phase flow is compressed air and distilled water. The air supplied by Ingersoll Rand T-30 Model 2545 compressor first passes through a regulator, filter, and lubricator circuit and then through a submerged helical coil heat exchanger. Next, the air is again passed through a filter and then fetched to Micro Motion Elite Series Model LMF 3M and CMF 025 Coriolis gas mass flow meters where the mass flow rate of air is controlled precisely using a Parker (24NS 82(A)-8LN-SS) needle valve. The compressed air is then allowed to enter the test section through a spiral static mixer. The liquid phase in the form of distilled water stored in a 50 gallon tank is circulated in the system using a Bell and Gossett (series 1535, model number 3445 D10) centrifugal pump. The distilled water is first passed through Aqua-pure AP12-T purifier and then through ITT model BCF 4063 shell and tube heat exchanger. The distilled water is then passed through an Emerson Coriolis mass flow meter (Micro Motion Elite Series model number CMF 100) where the mass flow rate of the liquid phase entering the test

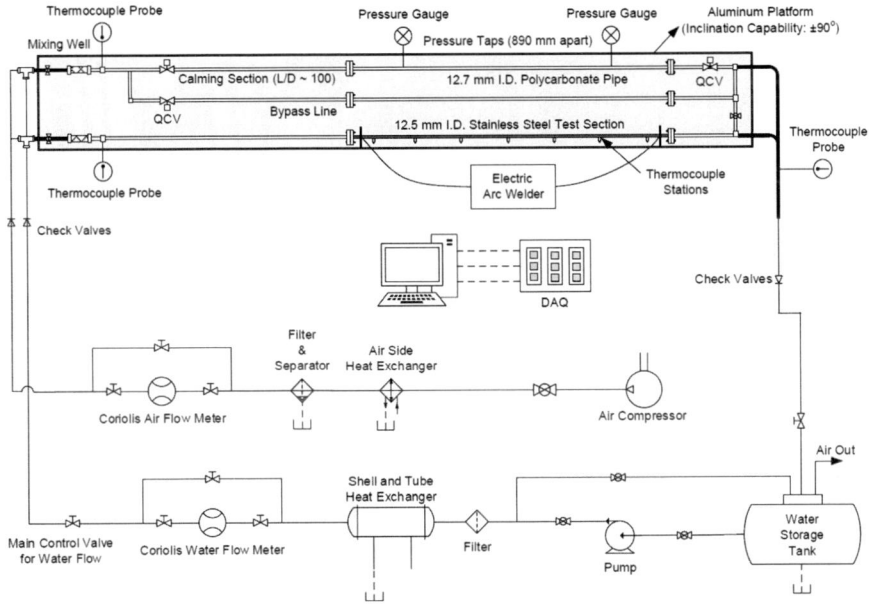

Fig. 2.1 Schematic of the two-phase flow experimental setup used for flow visualization and void fraction, pressure drop, and heat transfer measurements

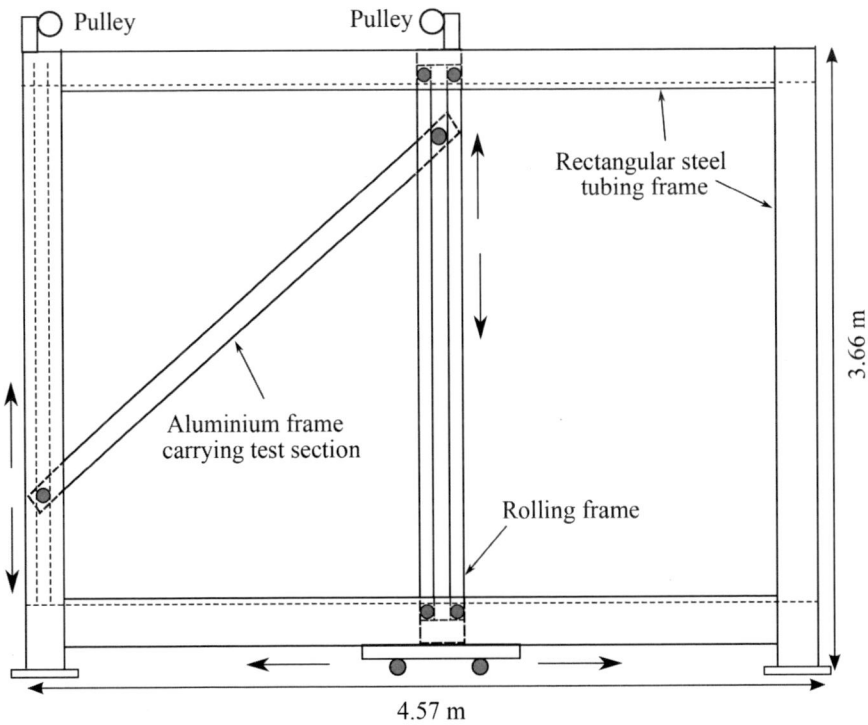

Fig. 2.2 Schematic of variable inclination frame carrying test section

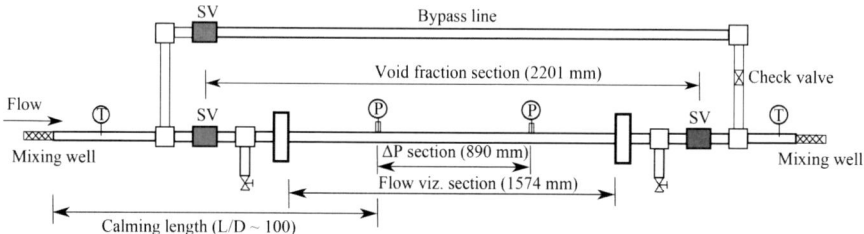

Fig. 2.3 Schematic of the details of the void fraction, pressure drop, and flow visualization test sections

section is controlled. Later, the water is allowed to mix with air in Koflo model 3/8-40C-4-3V-23/8 static mixer. The mixer is mounted right before the entrance to the test section. Two-phase flow measurements are carried out for all major flow patterns by systematically varying the gas flow rates for fixed values of liquid flow rates. The gas and liquid flow rates are varied in a range of 0.001–0.2 kg/min and 1–10 kg/min, respectively, to generate all possible flow patterns observed in horizontal and upward and downward inclined two-phase flow. For isothermal measurements (void fraction and pressure drop), system temperature is maintained between 20 °C and 25 °C, and the system pressure is found to vary between 1 and 3 bar. For these flow conditions, the Reynolds number for gas and liquid phase was found to vary between 100–19,000 and 2000–18,000, respectively. During heat transfer measurements, although the pipe wall temperature varied between 40 °C and 60 °C, the bulk temperature (T_b) of two-phase mixture was regulated between 17 °C and 25 °C using the shell and tube heat exchanger. Thus, the effect of system temperature on the superficial liquid phase Reynolds number (Re_{SL}) during heat transfer measurements was not considered significant. Consequently, Re_{SL} measured during heat transfer experiments was comparable to that for void fraction and pressure drop measurements. For each combination of gas and liquid flow rate, void fraction is measured using quick closing solenoid valves (SV) with a closing time of 0.1 s. When the valves are triggered, the two normally open valves close, and the normally closed valve opens simultaneously. In this manner, a two-phase sample is trapped in the void fraction section, while the air–water mixture is allowed to continue flowing through the bypass line (see Fig. 2.3). Air–water mixture trapped in the void fraction section is drained, and the volume of the liquid is measured. With the measured volume of the liquid phase (V_L), the void fraction (α_G) can be determined from the total volume of the test section (V_t) using $\alpha_G = 1 - V_L/V_t$. The uncertainty associated with the volume of the liquid is ± 0.5 cm^3, and the uncertainty associated with the total volume of the test section is ± 2 cm^3. Using the uncertainty propagation method prescribed by Kline and McClintock (1953), the uncertainty associated with measurements of void fraction is estimated to be between $\pm 1.25\%$ and $\pm 10\%$. The high percentage value of uncertainty is due to the small values of void fraction. Note that, in the transient region of the two-phase flow, void fraction measurements were repeated eight to ten times, and the average value was used as a representative void fraction.

The total two-phase pressure drop is measured using a DP15 variable reluctance Validyne pressure transducer having an accuracy of 0.25% of full-scale range of the diaphragm. The two pressure diaphragms with a range of 0.5 psi (DP-26) and 2 psi (DP-32) are used to measure total two-phase pressure drop, and hence the associated worst-case uncertainty with total two-phase pressure drop is $\approx \pm 35$ Pa/m. The upstream and downstream pressure taps are placed 890 mm apart with the upstream pressure tap placed at $\approx 100D$ from the entrance of the test section (see Fig. 2.3). Since the pressure taps are connected to the differential pressure transducer through flexible connecting lines, the measured value of total two-phase pressure drop given by Eq. (2.1) is corrected for the amount of liquid phase trapped in these lines at any given pipe orientation. For the non-boiling two-phase flow conditions at near atmospheric system pressures, the accelerational component is usually very small since there is no expansion of the two-phase mixture and/or change in two-phase flow quality due to phase change process. Based on the absolute pressure measured at upstream and downstream pressure taps, the maximum contribution of the accelerational component is found to be $\leq 1\%$ of the total two-phase pressure drop, and hence it is neglected in the calculation of frictional pressure drop. The hydrostatic pressure drop as a function of void fraction and phase densities is as expressed by Eq. (2.2). Assuming negligible contribution from the accelerational component of two-phase pressure drop, the frictional component of two-phase pressure drop is obtained by subtracting the hydrostatic component from total component of two-phase pressure drop as given by Eq. (2.3):

$$\left(\frac{dp}{dz}\right)_t = \left(\frac{dp}{dz}\right)_h + \left(\frac{dp}{dz}\right)_f + \left(\frac{dp}{dz}\right)_a \qquad (2.1)$$

$$\left(\frac{dp}{dz}\right)_h = [\rho_L(1 - \alpha_G) + \rho_G \alpha_G]g \sin\theta = \rho_{MG}g \sin\theta \qquad (2.2)$$

$$\left(\frac{dp}{dz}\right)_f = \left(\frac{dp}{dz}\right)_t - \left(\frac{dp}{dz}\right)_h \qquad (2.3)$$

In order to accomplish heat transfer measurements, a constant heat flux in a range of 7 kW/m^2 to 60 kW/m^2 is provided to the test section using Miller Maxtron 750 electric arc welder that has a capacity to produce steady output of currents up to 450 Amps. The electric input is delivered to the test section using 6.35-mm-thick copper plates soldered to the ends of test section. These plates completely encircled the test section to ensure uniform distribution of electric input. Also, 1.27-cm-thick phenolic resin boards are placed on outer sides of these copper plates to prevent loss of heat from the heated test section to the non-heated portions of the pipe. A thick

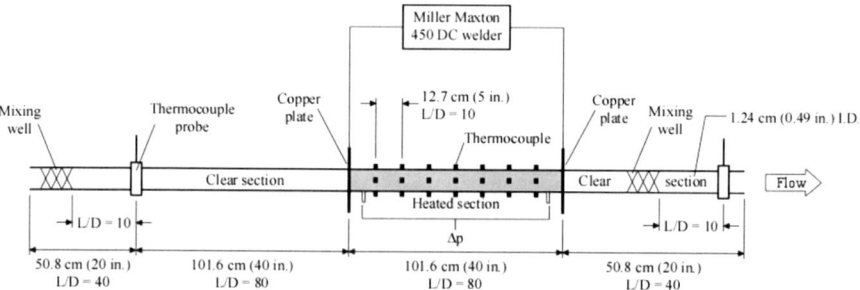

Fig. 2.4 Schematic of the details of the heat transfer test section

Fig. 2.5 Placement of the
thermocouples on the heat
transfer test section

Micro-Lok fiberglass insulator is used to provide insulation to ensure minimal heat
loss from the setup to the surrounding.

The two-phase inlet and outlet temperature is measured using Omega TMQSS-
06U-6 thermocouple probes. These thermocouple probes are inserted inside through
the pipe wall till they almost touch the other end of the pipe wall. This ensures that
the probes are always in contact with the two-phase mixture. CO1-T type thermocouples with an accuracy of $\pm 0.5\,°C$ are used to measure wall temperatures at seven
different stations spaced 12.7 cm apart along the 1016 mm ($\approx 80D$) long heated test
section. At each axial location, four thermocouples equally spaced over the pipe
circumference (top, bottom, and two side walls of the pipe) are cemented to the pipe
outer wall. Thus, the average values of pipe outer and inner wall temperature are
based on 28 temperature measurements along the test section length. See Figs. 2.4
and 2.5 for the details of the heat transfer test section and the placement of
thermocouples on the heat transfer test section.

The average values of inside wall temperature $(\overline{T}_{w,i})$, heat flux, and convective
heat transfer coefficient $(\overline{h})$ based on pipe outer wall temperature $(\overline{T}_{w,o})$ are
calculated using a finite difference formulation-based data reduction program developed by Ghajar and Kim (2006). The local (circumferential average at each axial
station) inner wall temperature $(\overline{T}_{w,i})$ and the associated heat transfer coefficient are
calculated using Eqs. (2.4) and (2.5), respectively. The two-phase convective heat
transfer coefficient (h_{TP}) is represented by the average of the measured local values
of $\overline{h}$ at each thermocouple station (N_{ST}) given by Eq. (2.6):

$$\overline{T}_{w,i} = \overline{T}_{w,o} + \frac{\dot{q}}{2\pi k \Delta z} \ln\left(\frac{D_o}{D_i}\right) \qquad (2.4)$$

$$\overline{h} = \frac{\dot{q}}{2\pi D \Delta z \left(\overline{T}_{w,i} - T_b\right)} \qquad (2.5)$$

$$h_{\mathrm{TP}} = \frac{1}{L}\int \overline{h}\,dz = \frac{1}{L}\sum_{j=1}^{N_{\mathrm{ST}}} \overline{h}_j\,\Delta z_j \qquad (2.6)$$

The uncertainty associated with measurements of two-phase heat transfer coefficient is determined using Kline and McClintock (1953) method. Higher experimental uncertainties (~25%) are found to be associated with the annular, intermittent, and stratified flow patterns. For slug and bubbly flow patterns, the average uncertainty associated with two-phase heat transfer coefficient is less than 10%. The higher values of uncertainty in measurement of h_{TP} are essentially due to two factors, namely, high heat balance error (difference between actual and enthalpy based heat transfer rates) and low temperature difference $\left(\overline{T}_{w,i} - T_b\right)$, i.e., the difference between average inner wall temperature and bulk temperature. The bulk temperature is the average of the temperatures measured at the inlet and exit of the test section. The heat balance error results from losses due to convection to the surrounding and/or storage within the test section itself. It is difficult to control the heat balance error however; it is found that uncertainty in measurement of h_{TP} could be reduced by maintaining at least a 4 °C temperature difference between pipe inner wall and mixture bulk temperature. Thus, the electrical input to the test section is adjusted to obtain $\left(\overline{T}_{w,i} - T_b\right) \geq 4$ °C such that the uncertainty in calculation of this temperature difference is not significantly propagated in calculation of two-phase heat transfer coefficient. The validity of pressure drop and heat transfer measurements carried out using this setup is checked by comparing the single-phase friction factor and heat transfer coefficient against the correlations of Churchill (1977) and Seider and Tate (1936). Measured values of single-phase data are found to be within $\pm 10\%$ of the values predicted by these correlations. For detailed information regarding experimental setup and uncertainties in measurement of different two-phase flow variables, please refer to the work of Cook (2008) and Bhagwat (2015).

Chapter 3
Flow Patterns, Flow Pattern Maps, and Flow Pattern Transition Models

Background

The physical structure of two-phase flow known as "flow pattern" is most fundamental among all two-phase flow variables and also forms the basis of two-phase flow modeling methods. Unlike single-phase flow theory, two-phase flow cannot be distinguished as laminar, transitional, or turbulent but rather is classified on the basis of several user-defined flow patterns.

Flow Patterns

The flow patterns observed in gas–liquid two-phase flow are quite intriguing in nature and depend on the alignment of one phase with respect to the other across the pipe cross section. The morphological variations in the structure of flow patterns are essentially due to the significantly different physical properties of the two phases such as viscosity and density, compressible nature of the gas phase, and the interaction between gravity, inertia, and buoyant forces. Correct understanding of the physical structure of flow patterns and their transition is instrumental in the general understanding of the mechanism of mass, momentum, and energy transfer in two-phase flow.

This preliminary section of the chapter is aimed at describing the physical structure of some key flow patterns observed in horizontal, upward, and downward vertical pipes as shown in Figs. 3.1, 3.2, and 3.3, respectively, and their physical variations as a function of varying two-phase flow conditions. The different flow patterns were generated by systematically varying gas and liquid flow rates in a range of 0.001–0.2 kg/min (Re_{SG}~100 to 18500) and 1–10 kg/min (Re_{SL}~2000 to 18000), respectively. Existence of a certain flow pattern was confirmed by visual

A. J. Ghajar, *Two-Phase Gas-Liquid Flow in Pipes with Different Orientations*, SpringerBriefs in Applied Sciences and Technology, https://doi.org/10.1007/978-3-030-41626-3_3

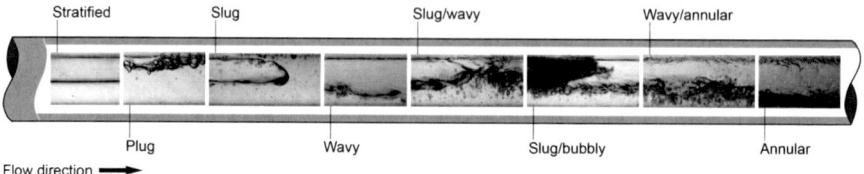

Stratified   Slug   Slug/wavy   Wavy/annular

Plug   Wavy   Slug/bubbly   Annular

Flow direction ➤

Fig. 3.1 Flow patterns in horizontal air–water two-phase flow. (Flow visualization done at the Two-Phase Flow Lab, OSU, Stillwater, OK)

Fig. 3.2 Flow patterns in upward vertical air–water two-phase flow. (Flow visualization done at the Two-Phase Flow Lab, OSU, Stillwater, OK)

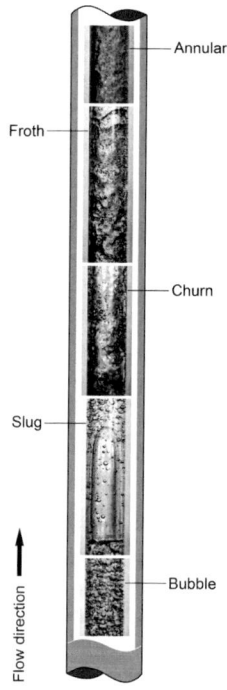

observation as well as through still photographs and videos recorded over a certain period of time using Nikon D3100 camera and 200 mm f/5.6 lens with a shutter speed of 1/4000 s. What follows is a brief description of two-phase flow patterns and their subtle morphological variations as a function of phase flow rates and pipe inclinations.

Bubbly Flow Bubbly flow pattern is observed at high liquid and low gas flow rates. Bubbly flow often called as dispersed bubbly flow is characterized by the flow of small gas bubbles dispersed in continuous liquid medium. The size (much smaller than the pipe diameter), shape (typically spherical or oblong), and distribution (uniform, center peaked, or wall peaked) of bubble density depend upon the pipe geometry and inclination, fluid properties, and phase flow rates. For horizontal and inclined two-phase flow (both upward and downward), the gas bubbles are always

Fig. 3.3 Flow patterns in
downward vertical air–water
two-phase flow. (Flow
visualization done at the
Two-Phase Flow Lab, OSU,
Stillwater, OK)

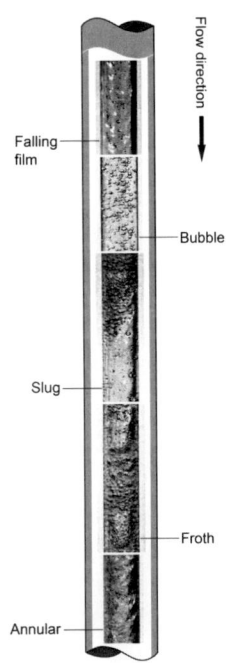

Fig. 3.4 Varying bubbly flow structures in horizontal, downward inclined, and vertical downward
two-phase flow. (Flow visualization done at the Two-Phase Flow Lab, OSU, Stillwater, OK)

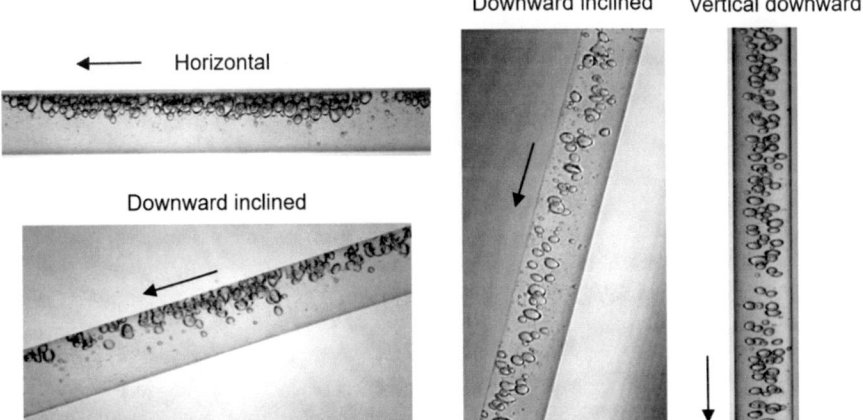

located near the pipe upper wall region (asymmetric flow) as shown in Fig. 3.4,
while for vertical flows (both upward and downward), the gas bubbles are evenly
distributed across the pipe cross section (symmetric flow); refer to Fig. 3.4. It is
observed that the gas and liquid flow rates significantly affect the size, shape, and
distribution of the gas bubbles dispersed in the liquid medium. In the case of vertical

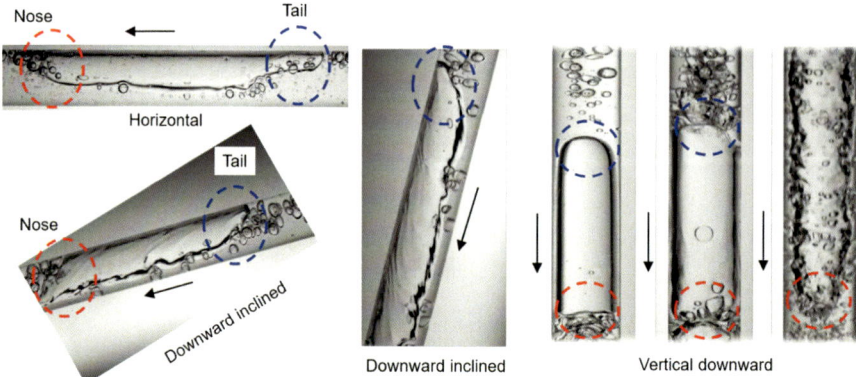

Fig. 3.5 Physical structure of slug flow pattern in horizontal, downward inclined, and vertical downward two-phase flow. (Flow visualization done at the Two-Phase Flow Lab, OSU, Stillwater, OK)

downward flow, at low gas flow rates, the gas bubbles appear to be distributed only in the central region of the pipe with single-phase liquid occupying the near wall region. This appearance of bubbly flow in vertical downward flow is essentially due to the "coring phenomenon" characterized by the repulsive force exerted by the pipe wall on the liquid phase. This type of flow behavior has been observed and reported by other investigators (Oshinowo 1971, Usui and Sato 1989, Yijun and Rezkallah 1993, Terekhov and Pakhomov 2008) for pipe diameters in a range of 9–45 mm. This coring flow behavior in vertical downward flow is evident from Fig. 3.4.

Slug Flow Slug flow is characterized by the alternate flow of gas and liquid slugs. The size (elongated and comparable to pipe diameter), shape (usually bullet shaped with hemispherical nose and a blunt tail), and frequency of slugs depend upon the pipe geometry, pipe orientation, and fluid properties. As shown in Fig. 3.5, for horizontal and non-vertical upward pipe inclinations, the gas slug is in the vicinity of the pipe upper wall, while it is symmetrically distributed around the pipe axis in vertical flow orientation. The translational velocity of the gas slug along the pipe axis is a function of pipe orientation such that the upward pipe inclinations aid the slug motion, while the downward pipe inclinations oppose the slug motion. The length of the gas slugs reduces, while their frequency increases with increase in the liquid flow rate. The opposite is true for a fixed liquid flow rate and decreasing gas flow rates.

In downward inclined flow (see Fig. 3.5), the buoyancy acting on the gas phase is in the direction opposite to that of the mean flow that causes distortion of the slug nose, and a tail pointing in the direction opposite to that of the flow direction is observed. It is also worthwhile to mention that the shape of the slug nose in downward inclinations is found to change with change in the phase flow rates depending upon the interaction between buoyant and inertia forces. In the case of vertical downward flow, at low liquid and gas flow rates, the elongated slug bubble appears very similar to that of the vertical upward flow, i.e., gas bubble nose appears to be pointed in the upward direction and remains undistorted. As shown in Fig. 3.5

with increase in the gas and liquid flow rates, the slug nose and tail (both ends of slug) become flat, and with further increase in phase flow rates, the slug nose points in the direction of the mean flow. These different types of slug nose in vertical downward flow are an indication of the different residence times of the gas phase in the test section (different magnitudes of interaction between buoyant and inertial forces) that influence the void fraction, pressure drop, and heat transfer characteristics of two-phase flow.

Intermittent Flow The intermittent two-phase flow that typically exists at moderate gas and liquid flow rates is a representation of two-phase flow having chaotic, pulsating, and indefinite phase alignment characteristics. Different subcategories of key flow patterns such as slug/wavy, slug/bubbly, and wavy/wavy annular shown in Fig. 3.1 for horizontal two-phase flow and churn and froth flow (see Figs. 3.2 and 3.3) exhibit disorderliness, intermittency, and indefinite flow structure, and hence it is appropriate to club all these different sub-flow patterns together to call as intermittent flow. This approach of calling certain flow patterns as intermittent flow has been adopted by many researchers to reduce the ambiguity in identifying the transition boundaries between key flow patterns. Previous works reported in the literature by Beggs and Brill (1973), Mukherjee (1979), and Barnea et al. (1982) have adopted similar flow pattern nomenclature for this type of flow behavior.

Stratified Flow Stratified flow exists in the form of gas and liquid layers flowing parallel to each other such that the pipe bottom wall is in contact with the liquid phase, while the upper wall is in contact with the gas phase. Stratified flow in general can be classified as *smooth stratified* and *wavy stratified* flow. Smooth stratified flow appears at low liquid and low gas flow rates and is characterized by the complete separation of the two phases sharing a smooth and stable interface, whereas, at high gas flow rates, the wavy stratified flow exists, and it is characterized by the wavy, dynamic, and rough gas–liquid interface. At relatively high gas and liquid flow rates, the interfacial instability generates disturbance waves at the interface that grow and tend to touch the pipe wall giving an appearance of rolling wave. This type of stratified flow is recognized as *rolling wave* flow; however, the general flow structure still resembles the stratified flow pattern. The flow visualization of the three different forms of stratified flow described above is presented in Fig. 3.6.

A specific case of stratified flow at low liquid and low to moderate gas flow rates is identified as "falling film flow." This is observed in vertical downward and near vertical downward pipe inclinations of two-phase flow. The falling film flow is characterized by the downward flow of thin liquid film flowing quiescently along the pipe wall, while the gas phase flows through the central region. Falling film flow at low liquid flow rates has a strong tendency of creating dry spots on the pipe wall and hence may be regarded as undesirable flow pattern for the non-boiling two-phase flow involving high heat flux at the pipe wall. The physical form of falling film flow for a fixed liquid flow rate and increasing gas flow rates is depicted in Fig. 3.6. The dry spots on the pipe wall surface at low gas flow rates are quite evident, and with increase in the gas flow rate, liquid droplets are observed to entrain the gas core. Due

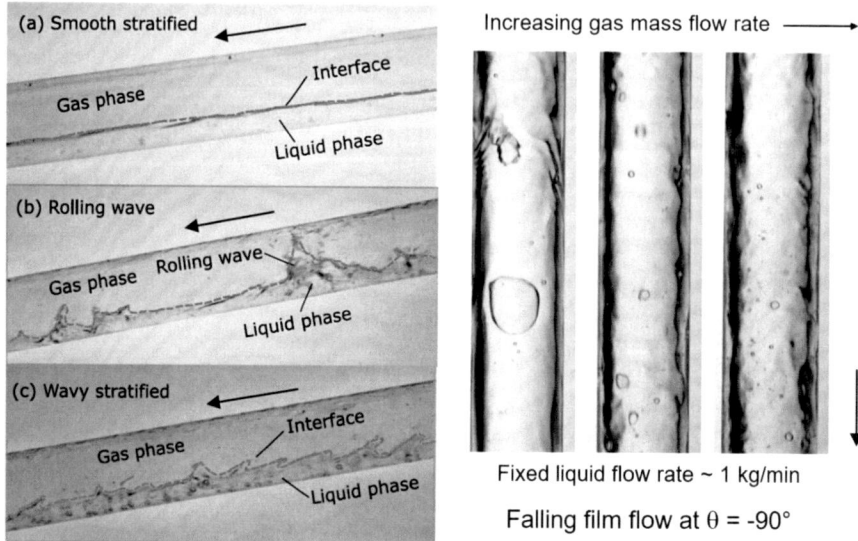

Fig. 3.6 Physical structure of stratified flow pattern in downward pipe inclinations and falling film flow in vertical downward flow. (Flow visualization done at the Two-Phase Flow Lab, OSU, Stillwater, OK)

to the physical resemblance of falling film with the annular flow, some researchers have classified this flow pattern as a subregion of annular flow. However, there is a significant difference in these two flow patterns in terms of the intensity of gas core turbulence, interfacial shear, thickness of liquid film, and the amount of liquid entrainment.

Annular Flow Annular flow appears in the form of a turbulent gas core surrounded by the thin/thick (depending upon liquid flow rates) and wavy liquid film. The liquid film distribution in annular flow is affected by the pipe orientation such that for horizontal flow, the asymmetric liquid film is thick at the pipe bottom while it is evenly distributed around the circumference in case of vertical two-phase flow; see Fig. 3.7. Based on the visual observations in Fig. 3.7, the distribution of film thickness around the pipe circumference (or film thickness at the pipe top and bottom wall for a 2D flow visualization) is found to depend on the gas and liquid phase flow rates and the pipe orientation. For a fixed liquid and gas flow rate, the effect of change in the pipe orientation from horizontal is to make the liquid film distribution more symmetric, whereas for a fixed pipe orientation and liquid flow rate, the effect of increase in the gas flow rate is to reduce the liquid film thickness in contact with the pipe wall. This effect of pipe inclination and gas flow rate on the circumferential distribution of liquid film thickness plays a pivotal role in the determination of the local as well as overall two-phase heat transfer coefficient.

Depending upon the gas and liquid flow rates, the gas-liquid interface may be stable or perturbed by small amplitude waves causing the liquid droplets to detach

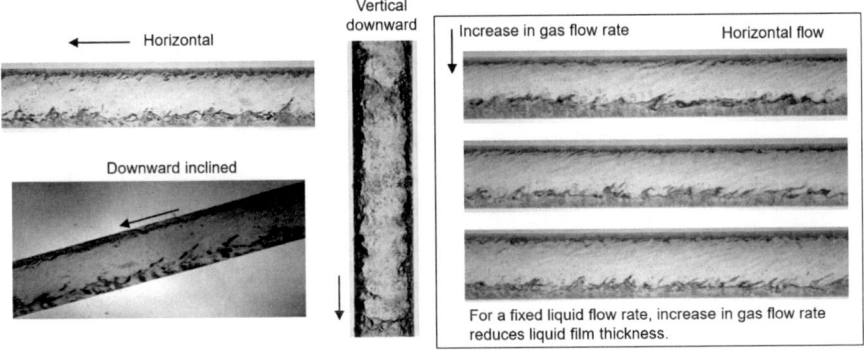

Fig. 3.7 Annular flow in horizontal, downward inclined, and vertical downward two-phase flow. (Flow visualization done at the Two-Phase Flow Lab, OSU, Stillwater, OK)

from liquid film and enter the turbulent gas core known as the liquid entrainment phenomenon. The amplitude and frequency of interfacial disturbance waves and the amount of liquid entrainment depend on the phase flow rates, fluid properties, pipe geometry, and pipe inclination. A special form of annular flow also known as annular-mist flow exists when most of the liquid phase initially flowing in the form of film is entrained into the central gas core by the shearing action of the fast-moving gas core. This type of flow pattern is mostly encountered in applications involving high system pressure and high heat flux conditions.

Flow Pattern Maps

The flow pattern maps or flow maps serve as a tool to estimate the span and sequence of the appearance of different flow patterns with change in gas and liquid flow rates for a given set of flow conditions. The flow patterns and their transitions are defined qualitatively based on visual observations, and hence the accurate mapping of the transition between different flow patterns is highly subjective and totally depends upon the observer's perception. Moreover, the transition between different flow patterns is gradual and sensitive to several parameters such as pipe diameter, pipe orientation, and fluid properties making it extremely difficult to have a universal flow pattern map that can correctly predict the existence of different flow patterns covering a wide range of two-phase flow conditions. Although some researchers have attempted to develop quantitative methods such as the probabilistic flow regime mapping and power spectral density analysis of the acquired pressure signal to predict the existence of certain flow patterns, these methods are not universal and are mostly limited to certain flow conditions. Moreover, these methods also rely upon prior information of visual inspection of flow patterns.

Despite the lack of the universal flow pattern map, some of the existing most referred to flow pattern maps are those of Mandhane et al. (1974) and Taitel and

Dukler (1976) for horizontal two-phase flow and Hewitt and Roberts (1969) for vertical upward flow. These flow pattern maps provide an idea about the sequence, extent, and transitions between different flow patterns and serve as a guideline in understanding the appearance of flow patterns as a function of gas and liquid flow rates. In general, keeping in mind the uncertainty, sensitivity, and qualitative identification of flow patterns, the following facts and limitations about the flow pattern maps must be perceived: (a) it is difficult to have a universal flow pattern map that can accurately predict the transition of one flow pattern to another for a wide range of two-phase flow conditions; (b) unlike represented in most flow pattern maps, the transition from one flow pattern to another is always gradual and cannot be presented in the form of a thin continuous line; and (c) correct identification of a flow pattern in the vicinity of the transition line depends on the judgment of an individual. For more discussion on different existing flow pattern maps, refer to Ghajar and Bhagwat (2017).

The purpose of this section is to present an overview of the effect of pipe orientation on the transition between different two-phase flow patterns. The flow pattern maps for different pipe orientations are presented with reference to the horizontal flow pattern map. It should be mentioned that the flow pattern maps presented in this section were developed for two-phase flow of air–water through a 12.7 mm I.D. pipe at different orientations and are applicable for similar flow conditions. The flow pattern maps for each orientation in upward and downward inclinations are presented in Figs. 3.8, 3.9, 3.10, and 3.11. These flow pattern maps are based on the experimental observations and data collected at the Two-Phase Flow Lab, Oklahoma State University (OSU).

In the case of both upward (Fig. 3.8) and downward (Fig. 3.10) inclined two-phase flow, regardless of the pipe orientation, the bubbly flow occurs for low values of gas and high values of liquid mass flow rates, whereas the annular flow exists for low liquid and high values of gas mass flow rates. In the case of upward inclined two-phase flow, slug flow exists at low to moderate liquid flow rates and low gas flow rates. The boundary between slug and intermittent flow is influenced by the change in pipe orientation. It is found that the increase in pipe orientation from horizontal causes this transition boundary to shift toward lower gas flow rates. This is possibly because increase in the pipe orientation assists buoyancy and facilitates formation (agglomeration of gas bubbles) and motion of gas slugs.

The transition between intermittent and annular flow regime is virtually unaffected by the pipe orientation due to the fact that this is a shear-driven flow regime where the pipe orientation has a little effect on the physical structure of this flow pattern. However, during intermittent flow regime, an early transition from wavy slug to wavy annular flow could be observed with increase in the upward pipe inclination. The transition boundary between bubbly and slug flow is also found to be affected by the change in pipe orientation. The increase in pipe orientation from horizontal to upward direction causes early transition from slug to bubbly flow. The transition between bubbly-slug, slug-intermittent, and intermittent-annular is evident from Fig. 3.9. In the case of horizontal flow at low liquid flow rates, a short region of intermittent flow exists in the transition from stratified to annular. The intermittent

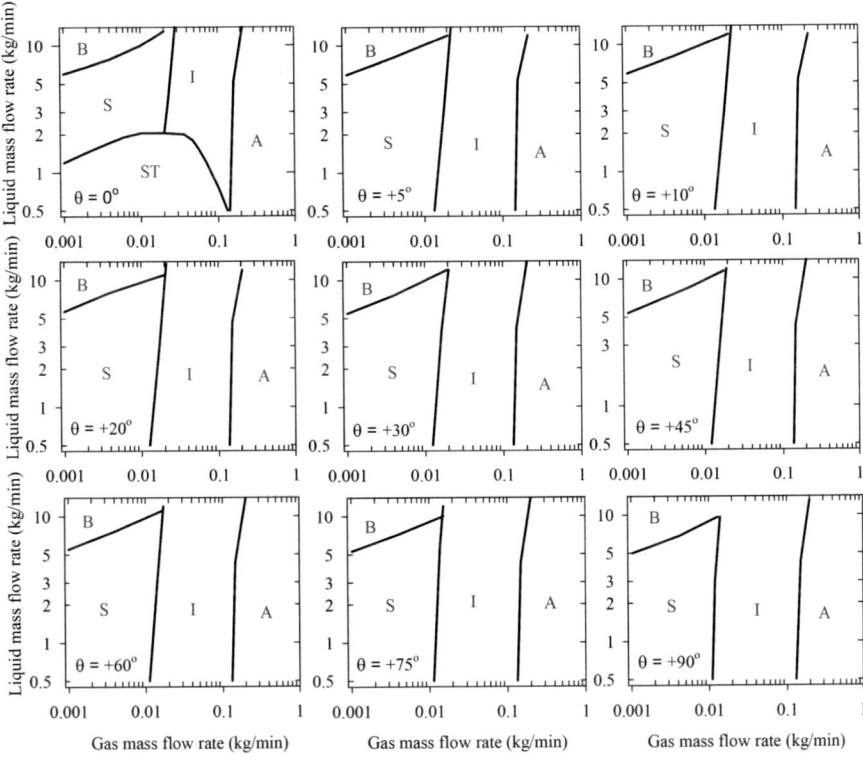

Fig. 3.8 Flow pattern map for upward inclined two-phase flow. (Based on flow visualization done at the Two-Phase Flow Lab, OSU, Stillwater, OK.) (B bubbly, S slug, I intermittent, A annular)

flow pattern identified in this narrow range of gas flow rates is mostly wavy annular in nature. At low liquid flow rates approximately 0.5 kg/min, stratified flow directly transits to annular flow without entering into intermittent flow regime.

In the case of downward inclined two-phase flow (Fig. 3.10), the transition between stratified-slug and slug-intermittent flow regimes is seen to be significantly affected by the change in pipe orientation. As mentioned earlier, the stratified flow appears for low liquid and low to moderate gas flow rates and exists for horizontal and all downward inclinations considered in this study. It is found that the increase in downward pipe inclination expands the region occupied by the stratified flow on the flow pattern map toward higher liquid flow rates and at the expense of slug and intermittent flow regimes.

It is seen from Fig. 3.11 that for low gas flow rates ($\dot{m}_G < 0.01$ kg/ min), the transition boundary between stratified and slug flow consistently shifts toward higher liquid flow rates with increase in the gas flow rate up to $-45°$ from horizontal. With further increase in downward pipe inclination, the transition line between stratified and slug flow shifts back to lower liquid flow rates. For lower gas flow rates, it is found that the stratified flow exists for liquid flow rates as high as

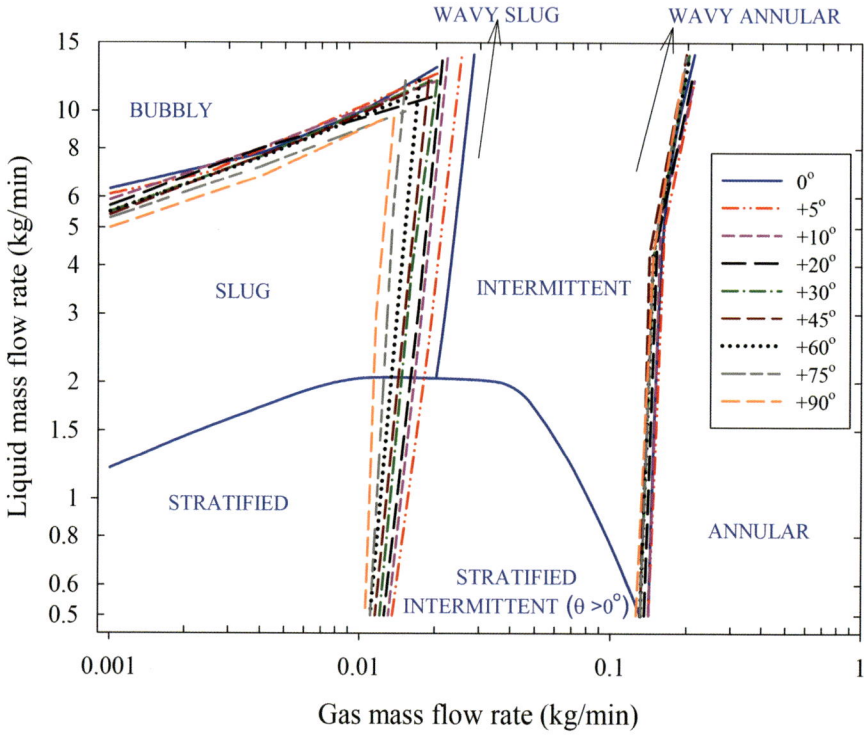

Fig. 3.9 Combined flow pattern map for upward inclined pipe orientations. (Based on flow visualization done at the Two-Phase Flow Lab, OSU, Stillwater, OK)

5.5 kg/min for −45° pipe orientation. With further increase in pipe orientation at −60°, the stratified flow cannot sustain beyond liquid flow rate of 4 kg/min. For moderate gas flow rates $(0.01 \leq \dot{m}_G \leq 0.05 \ \text{kg/min})$, the transition between stratified and intermittent flow regime is very gradual (as a function of liquid flow rate and relatively independent of gas flow rate) until a threshold value of the gas flow rate is attained where the gas-liquid surface becomes significantly unstable and the liquid phase is splashed frequently on the pipe top wall. This is the point where the flow pattern gradually shifts to intermittent (wavy annular) flow and the transition line between stratified and intermittent flow patterns is steep and very sensitive to the change in gas and liquid flow rates. Figure 3.11 shows that this threshold value of gas mass flow rate is a function of pipe orientation, i.e., it increases until $\theta = -45°$ and then decreases thereafter. Note that for $\theta = -75°$, falling film flow coexists with stratified flow at low liquid and low to moderate gas flow rates. At this near vertical pipe orientation, the edges of gas–liquid interface climb up the tube periphery such that it occupies most of the pipe circumference. With increase in liquid flow rate, a thin film of liquid is observed at the pipe upper wall, and falling film flow is said to exist. As mentioned earlier, falling film flow is a special case of stratified flow for vertical and near vertical downward pipe orientations, and hence no distinction is

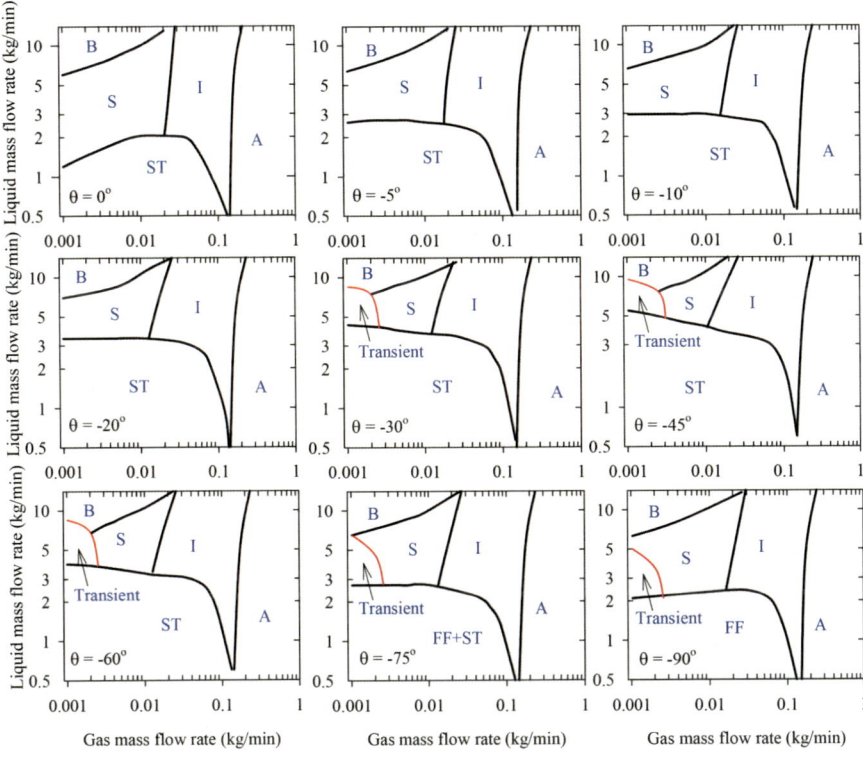

Fig. 3.10 Flow pattern map for downward inclined two-phase flow. (Based on flow visualization done at the Two-Phase Flow Lab, OSU, Stillwater, OK.) (B bubbly, S slug, I intermittent, ST stratified, A annular, FF falling film)

made between these two-phase flow patterns on the flow map in Fig. 3.10. For vertical downward pipe inclination ($\theta = -90°$), stratified flow is always in the form of falling film due to flow symmetry.

In the case of transition between slug and intermittent flows, with increase in the downward pipe inclination, the transition boundary is found to shift toward lower gas flow rates up to $\theta = -45°$ and then slightly moves toward higher gas flow rates thereafter. Similar to upward inclined flows, the transition between intermittent and annular flows is virtually not affected by the change in the pipe orientation. For very low liquid flow rates $\dot{m}_L < 1 \, \text{kg}/\min$, it is found that the flow pattern shifts directly from stratified to annular without entering the intermittent flow regime. Typically for $\dot{m}_L < 2 \, \text{kg}/\min$, the intermittent region is very narrow and is wavy annular in nature.

Another interesting characteristic of gas–liquid two-phase flow observed in steeper downward pipe inclinations is the transient behavior of the flow patterns. As graphically illustrated in Fig. 3.12, this transient behavior results in simultaneous or sequential existence of multiple flow patterns (usually stratified, slug, and bubbly

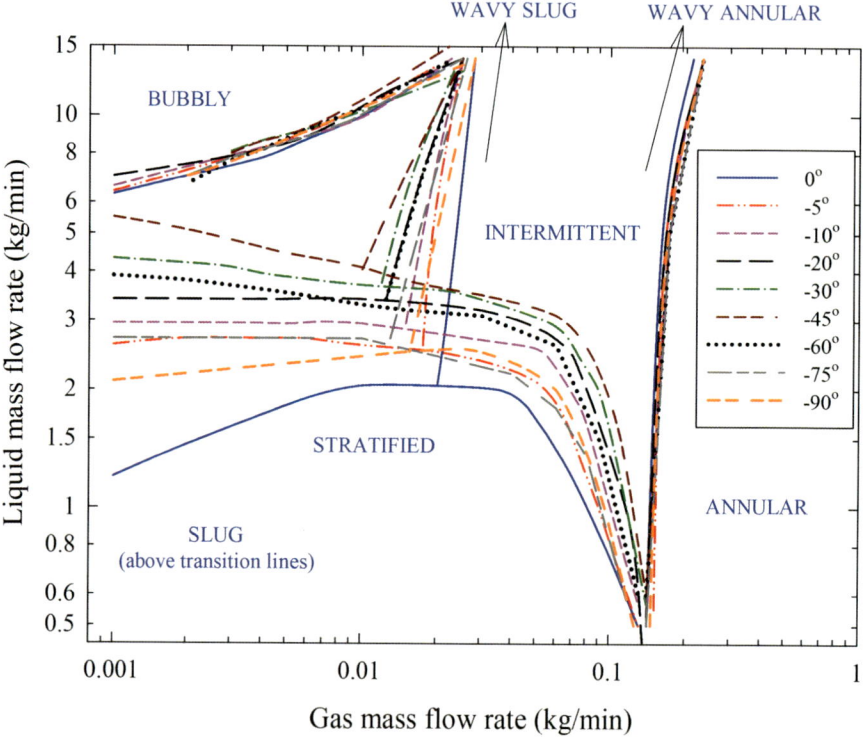

Fig. 3.11 Combined flow pattern map for downward inclined pipe orientations. (Based on flow visualization done at the Two-Phase Flow Lab, OSU, Stillwater, OK)

Fig. 3.12 Schematic of the general appearance of flow patterns during transient behavior of downward inclined gas–liquid two-phase flow. (Based on flow visualization done at the Two-Phase Flow Lab, OSU, Stillwater, OK)

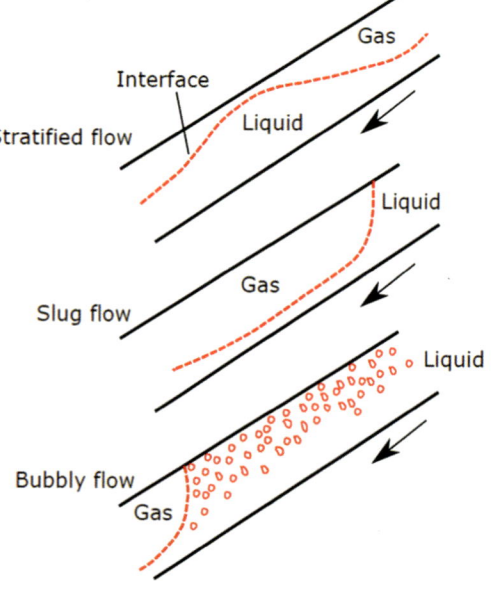

flows) in the test section. This phenomenon found in steeper downward pipe inclinations is verified visually as well as through the two-phase pressure drop signal recorded over a period of time as shown in Fig. 3.13. Comparison of visual observations with the pressure drop signal shows that the flat portion of the signal corresponds to the stratified flow, while the spiked portion of the signal toward negative values indicates the presence of bubbly and/or slug flow patterns. Note that the total pressure drop is negative because of higher negative hydrostatic head compared to positive frictional head in downward pipe inclinations. The pressure drop signal over a period of time confirms that these flow patterns (stratified/bubbly/slug) repeat one after another continuously over a period of time without establishing one single flow pattern. However, with increase in the gas flow rate, the time period between the occurrences of multiple flow patterns decreases. The flow pattern map reported in Fig. 3.10 shows that with increase in the gas flow rate, the transient behavior gradually diminishes, and the slug/intermittent flow behavior of the two-phase flow is restored. A plausible explanation for transient behavior of flow patterns can be given considering the two-phase flow physics in downward pipe inclinations. Near the stratified/non-stratified transition line, the downward moving gas–liquid interface is unstable. The unstable interface frequently splashes liquid on the top wall of the pipe causing frequent bridging of pipe cross section. Thus, the elongated gas slug gets trapped in the test section that appears to remain virtually stationary (hence also referred to as a pseudo-stationary slug) under the influence of buoyancy force acting on the gas phase in a direction opposite to that of the mean flow. The two-phase mixture coming from the upstream direction is accumulated over this pseudo-stationary slug, which further slips past the elongated gas bubble and eventually forces the gas bubble to move in the downstream direction. During this process, the liquid phase churns the elongated gas bubble causing it to disintegrate and move further in the form of discrete bubbles. After the gas pocket is pushed downstream, the incoming two-phase flow is again in the form of stratified flow pattern, and this process repeats continuously showing stratified, slug, and bubbly flow patterns periodically. A change in flow pattern from one to another (with significantly different flow structure) causes acceleration and deceleration of the two-phase mixture that results in pressure loss and pressure recovery in the two-phase flow system and hence results in an abrupt change in the pressure drop signal. It is found that this transient nature of two-phase flow aggravates with increase in downward pipe inclination and propagates toward slightly higher gas flow rates. At low gas flow rates, the time occupied by the gas slugs in the test section between two consecutive stratified flow appearances is higher than that at higher gas flow rates. Thus, it is clear that this transient behavior is undesirable and must be avoided in practical applications involving downward inclined gas–liquid two-phase flow.

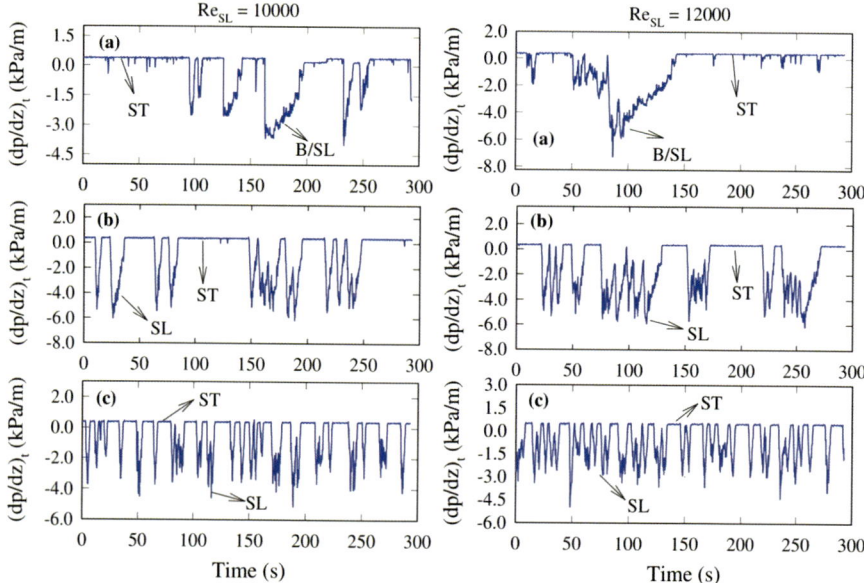

Fig. 3.13 Pressure drop signal to identify the transient behavior in downward inclined two-phase flow $\theta = -45°$, (**a**) $Re_{SG} = 180$, (**b**) $Re_{SG} = 360$, and (**c**) $Re_{SG} = 450$ (B bubbly, SL slug, and ST stratified). (Based on the data collected at the Two-Phase Flow Lab, OSU, Stillwater, OK)

Flow Pattern Transitions

Two-phase flow literature reports both mechanistic and empirical models to predict transition between different flow patterns. However, these transition models are developed based on the visual observations and limited data bank and are based on several simplifying assumptions that render uncertainty in their practical applications specifically if the two-phase flow under consideration is on the verge of transition. The following section provides some of the existing flow pattern transition models available in the two-phase flow literature. Considering the vague definition of intermittent flow pattern, the transition models listed here are only for well-defined flow regimes such as bubbly, slug, stratified, and annular flows.

Transition Between Bubbly and Slug Flows The transition from bubbly to slug flow occurs when bubbles no longer move independently and tend to coalesce to form a larger bubble usually known as Taylor bubble. The transition from bubbly to slug flow can be achieved with decrease in liquid flow rate at a constant gas flow rate or with increase in gas flow rate at a constant liquid flow rate. In general, two-phase flow literature shows that this transition from bubbly to slug flow occurs for void fraction in a range of 0.25–0.3. The transition from bubbly to slug flow will occur if Eq. (3.1) suggested by Taitel et al. (1980) is satisfied. This relationship is obtained by assuming that the maximum void fraction during transition from bubbly to slug flow is approximately 0.25 and that the bubble velocity is equal to the slip velocity

$(U_b \cong U_G - U_L)$. The bubble velocity at different pipe inclinations represented by Eq. (3.2) is the modified form of Haramathy (1960) equation for bubble rise velocity suggested by Taitel et al. (1980) and Barnea (1987):

$$U_{SL} \leq 3U_{SG} - 1.32\left(\frac{\sigma g(\rho_L - \rho_G)}{\rho_L^2}\right)^{0.25} \sin\theta \left.\right\} \quad (+60^\circ \leq \theta \leq +90^\circ) \quad (3.1)$$

$$U_b = 1.53\left[\frac{g\sigma(\rho_L - \rho_G)}{\rho_L^2}\right]^{0.25}\sqrt{1 - \alpha_G}\sin\theta \quad (3.2)$$

Note that Eq. (3.1) is recommended for use for near vertical pipe orientations $(+60^\circ \leq \theta \leq +90^\circ)$ since as per Taitel et al. (1980) and Barnea (1987), it is difficult for bubbly flow to exist in horizontal and near horizontal pipe orientations. Moreover, Taitel et al. (1980) also reported that the bubbly flow would exist if the pipe diameter of two-phase flow under consideration follows Eq. (3.3):

$$D > 19\left(\frac{(\rho_L - \rho_G)\sigma}{\rho_L^2 g}\right)^{0.5} \quad (3.3)$$

Following a similar concept, Mishima and Ishii (1984) presented a model to predict the transition between bubbly and slug flow (valid for vertical upward flow) of the form shown in Eq. (3.4). The transition from bubbly to slug flow would occur if Eq. (3.4) is satisfied. The parameter C_o is expressed as $C_o = 1.2 - 0.2\sqrt{\rho_G/\rho_L}$ such that for near atmospheric operating conditions $C_o \approx 1.2$. The bubble velocity equation (Eq. 3.5) used by Mishima and Ishii (1984) is similar to Eq. (3.2), however with different multiplying factors and exponents:

$$U_{SL} \leq \left(\frac{3.33}{C_o} - 1\right)U_{SG} - \frac{0.76}{C_o}\left(\frac{\sigma g\Delta\rho}{\rho_L^2}\right)^{0.25} \left.\right\} \quad (\theta = +90^\circ) \quad (3.4)$$

$$U_b = 1.41\left[\frac{g\sigma(\rho_L - \rho_G)}{\rho_L^2}\right]^{0.25}(1 - \alpha_G)^{1.75} \quad (3.5)$$

For vertical upward flow, a comparison between Eqs. (3.1) and (3.4) shows that transition from bubbly to slug flow will occur at relatively lower U_{SL} using the criteria of Mishima and Ishii (1984). This discrepancy is essentially due to their assumption of $\alpha_G \approx 0.3$ during the transition and a slightly different form of bubble velocity (U_b) used by Mishima and Ishii (1984). As mentioned earlier, bubbly flow pattern can be further classified as bubbly-slug and dispersed bubbly flow. Thus, the bubbly flow predicted by Eqs. (3.1) and (3.4) may contain both bubbly-slug and dispersed bubbly two-phase flow. Unlike bubbly (elongated bubbly/slug bubbly) flow, the dispersed form of bubbly flow is known to exist in the entire range of pipe orientations, and its existence could be determined using the transition model of

Barnea (1986). The model of Barnea (1986, 1987) requires calculation of three different diameters, namely, maximum bubble diameter (d_{max}), critical bubble diameter above which bubble is deformed (d_{def}), and critical bubble diameter below which migration of bubbles is prevented (d_{migr}). These different bubble diameters could be determined from Eqs. (3.6, 3.7, and 3.8), respectively. Thus according to Barnea (1986), dispersed bubbly flow would exist if both $d_{max} < d_{def}$ and $d_{max} < d_{migr}$ conditions are satisfied. For these conditions typically at high liquid flow rates, dispersed bubbly flow exists since the turbulent fluctuations are capable to break down bubbles to spherical shape and also suppress the bubble coalesce that leads to slug formation:

$$d_{max} = \left[0.725 + 4.15\left(\frac{U_{SG}}{U_M}\right)^{0.5}\right]\left(\frac{\sigma}{\rho_L}\right)^{0.6}\left(\frac{2f_M U_M^3}{D}\right)^{-0.4} \tag{3.6}$$

$$d_{def} = 2\left[\frac{0.4\sigma}{(\rho_L - \rho_G)g}\right]^{0.5} \tag{3.7}$$

$$d_{migr} = 0.375\left(\frac{f_M U_M^2 \rho_L}{g\cos\theta(\rho_L - \rho_G)}\right) \tag{3.8}$$

In Eq. (3.6) f_M is friction factor based on two-phase mixture velocity U_M and can be calculated using any single-phase friction factor correlation. Considering the packing density and packing configuration of bubbles, Barnea (1986) showed that this transition model for dispersed bubbly flow is valid for void fraction in a range of $0 < \alpha_G \leq 0.52$. For dispersed bubbly flow, the slippage at gas-liquid interface is quite small, and it can be assumed that $\alpha_G \approx \lambda$. Thus, the transition from dispersed bubbly to slug or intermittent flow would occur at $\lambda = U_{SG}/U_M > 0.52$.

Transition Between Stratified and Non-stratified Flows As established earlier, the transition boundary between stratified-slug and stratified-annular flow is very sensitive to the change in downward pipe inclination. Considering the dependency and practical difficulty in the modeling of two-phase flow variables such as void fraction, pressure drop, and non-boiling two-phase heat transfer in stratified two-phase flow, it is desirable to have a stratified flow pattern-specific model to predict these parameters. As such, it is first required to be able to correctly predict the existence of stratified flow as a function of phase flow rates, fluid properties, and pipe orientation. This section presents a brief review of the mechanistic flow pattern transition model of Taitel and Dukler (1976) and an empirical model of Bhagwat and Ghajar (2015a) to predict the transition to stratified flow in horizontal and downward pipe inclinations.

The most referred transition model in two-phase flow literature to separate out stratified flow from non-stratified flow is that of Taitel and Dukler (1976). Their transition model is based on the momentum balance between the two phases, assumes a smooth and flat gas-liquid interface such that $f_i/f_G \approx 1$, and gives the

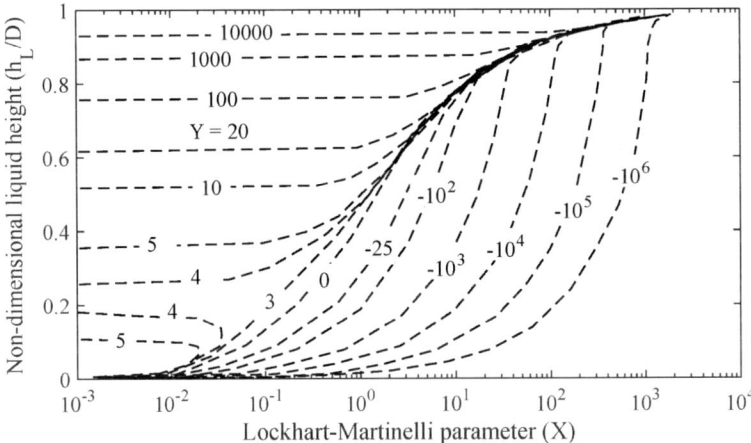

Fig. 3.14 Graphical solution to determine liquid-level height in stratified flow. (Adapted from Taitel and Dukler 1976)

transition line equation that requires use of a graphical solution. As per the model of Taitel and Dukler (1976), stratified flow will exist if Eq. (3.9) is satisfied where F is defined by Eq. (3.10):

$$F^2\left[\frac{1}{\left(1-\tilde{h}_L\right)^2}\frac{\tilde{U}_G\sqrt{1-\left(2\tilde{h}_L-1\right)^2}}{\tilde{A}_G}\right]<1 \qquad (3.9)$$

$$F=\sqrt{\frac{\rho_G}{\rho_L-\rho_G}}\frac{U_{SG}}{\sqrt{gD}\cos\theta} \qquad (3.10)$$

The non-dimensional parameters $\tilde{A}_G$ and $\tilde{U}_G=(\pi/4)/\tilde{A}_G$ are functions of non-dimensional liquid height $\left(\tilde{h}_L=h_L/D\right)$. The non-dimensional pipe cross-sectional area occupied by the gas phase is calculated using Eq. (3.11). As shown in Fig. 3.14, the non-dimensional liquid height $\tilde{h}_L$ is obtained through the graphical solution of $\tilde{h}_L$ vs. X for different values of Y. The parameters X and Y are obtained from Eqs. (3.12) and (3.13), respectively:

$$\tilde{A}_G=0.25\left[\cos^{-1}\left(2\tilde{h}_L-1\right)-\left(2\tilde{h}_L-1\right)\sqrt{1-\left(2\tilde{h}_L-1\right)^2}\right] \qquad (3.11)$$

$$X = \sqrt{\frac{(dp/dz)_{\mathrm{L}}}{(dp/dz)_{\mathrm{G}}}} \qquad (3.12)$$

$$Y = \frac{(\rho_{\mathrm{L}} - \rho_{\mathrm{G}})g \sin \theta}{(dp/dz)_{\mathrm{G}}} \qquad (3.13)$$

The application of Taitel and Dukler (1976) transition model to predict stratified flow is rather tedious due to the use of a graphical solution to determine non-dimensional liquid height $\tilde{h}_{\mathrm{L}}$ and consequently other parameters such as $\tilde{U}_{\mathrm{G}}$ and $\tilde{A}_{\mathrm{G}}$. The use of a graphical solution to determine h_{L}/D could be circumvented by adopting an iterative solution technique. For more details refer to Taitel and Dukler (1976).

Taitel and Dukler (1976) further attempted to classify stratified flow as smooth stratified and wavy stratified flows. They suggested that the transition between smooth and wavy stratified flow is associated with phenomenon of wave generation at the gas–liquid interface and that the smooth stratified flow exists for superficial gas velocities (U_{SG}) that satisfy Eq. (3.14). In practice, smooth stratified flow exists for a very narrow range of two-phase flow conditions and is of less practical significance compared to other flow patterns.

$$U_{\mathrm{SG,t}} \geq \left(\frac{4\nu_{\mathrm{L}}g\Delta\rho \cos \theta}{0.01\rho_{\mathrm{L}}U_{\mathrm{L}}} \right)^{0.5} \qquad (3.14)$$

A simplifying approximation to the Taitel and Dukler (1976) model has been presented by Cheng et al. (1988). They approximated the trend of Eq. (3.9) and presented a criterion given by Eq. (3.15) to identify the existence of stratified flow pattern in horizontal pipes in terms of Lockhart and Martinelli (1949) parameter applicable for turbulent–turbulent gas–liquid two-phase flow where Fr is the modified Froude number given by Eq. (3.16):

$$Fr_{\mathrm{SG}} \leq \left(\frac{1}{0.65 + 1.11X_{\mathrm{tt}}^{0.6}} \right)^{2} \qquad (3.15)$$

$$Fr_{\mathrm{SG}} = \frac{Gx}{\sqrt{gD\rho_{\mathrm{G}}(\rho_{\mathrm{L}} - \rho_{\mathrm{G}})}} = \frac{U_{\mathrm{SG}}}{\sqrt{gD}} \sqrt{\frac{\rho_{\mathrm{G}}}{\rho_{\mathrm{L}} - \rho_{\mathrm{G}}}} \qquad (3.16)$$

Combining Eqs. (3.15) and (3.16) gives the criteria for existence of stratified flow in terms of two-phase mixture mass flux. Equation (3.17) represents the maximum value of two-phase mixture mass flux (for a given quality and fluid properties) below which stratified flow will always exist and is recommended for use only in the case of horizontal two-phase flow:

$$G_{max} = \frac{\sqrt{g D \rho_G (\rho_L - \rho_G)}}{x} \left(\frac{1}{0.65 + 1.11 X_{tt}^{0.6}} \right)^2 \Bigg\} \quad (\theta = 0°) \qquad (3.17)$$

Using the concept of growth of disturbance waves on the gas-liquid interface of stratified flow, Mishima and Ishii (1980) presented a criterion to determine the limit of stratified flow in horizontal two-phase flow as given by Eq. (3.18). Note that use of this correlation also requires determination of liquid height (h_L) which could be obtained using graphical solution given by Fig. 3.14:

$$U_G - U_L = 0.487 \sqrt{\frac{g(D_h - h_L)(\rho_L - \rho_G)}{\rho_G}} \Bigg\} \quad (\theta = 0°) \qquad (3.18)$$

The recent work of Bhagwat and Ghajar (2015a) expressed by Eq. (3.19) provides an empirical correlation to predict the existence of stratified flow in horizontal and downward pipe inclinations. Similar to the approach of Cheng et al. (1988) and Bhagwat and Ghajar (2015a), correlation attempts to approximate the trends of Taitel and Dukler (1976) in terms of non-dimensional parameters of Fr_{SG} and X and is developed based on the experimental data in horizontal and downward pipe inclinations. The Froude number in Eq. (3.19) is similar to that given by Cheng et al. (1988) in Eq. (3.16). The variables C_1, C_2, C_3, and C_4 required to solve Eq. (3.19) are expressed by Eqs. (3.20, 3.21, 3.22, and 3.23). The non-dimensional pipe diameter (D^+) is the normalized pipe diameter defined by $D^+ = D/0.0254$:

$$Fr_{SG} \leq (0.6 + C_2) \exp\left(-C_1 C_2 X^{C_3}\right) X^{-C_4} \left. \begin{array}{l} \\ \\ \\ \\ \\ \end{array} \right\} \quad \begin{array}{l} D = 8.9 - 300 \ mm \\ \rho_L = 780 - 1420 \ kg/m^3 \\ \rho_G = 1.2 - 35 \ kg/m^3 \\ \theta = 0° \ to - 90° \\ \mu_L = 0.0002 - 0.08 \ Pa \ s \end{array}$$

$$(3.19)$$

$$C_1 = 1.3 \ln\left(D^+\right) + 2.5 \qquad (3.20)$$

$$C_2 = \frac{C_4^{0.65}}{[1 + 2 \sin(2|\theta|) \times (1 + 10 \tanh(1/|\theta|))]} \qquad (3.21)$$

$$C_3 = \begin{cases} 0.65(D^+)^{-0.15}((\rho_L - \rho_G)/1000) & : \rho_L < 1000 \ kg/m^3 \\ 0.65(D^+)^{-0.15} & : \rho_L \geq 1000 \ kg/m^3 \end{cases} \qquad (3.22)$$

$$C_4 = 0.2\sqrt{1/D^+} \tag{3.23}$$

Variables C_1 and C_2 account for the combined effect of pipe diameter and pipe orientation on the shift in transition line. The shift in the transition line by including these variables is in accordance with the observations of Nguyen (1975), Barnea et al. (1982), Crawford et al. (1985), and Ghajar and Bhagwat (2014a) that the increase in downward pipe inclination shifts the transition between stratified and non-stratified flow pattern toward higher values of liquid flow rates or alternatively toward higher values of X. Variable C_3 ensures that the transition line between stratified and non-stratified flow is shifted toward higher values of X with a decrease in liquid phase density. The variables C_1, C_2, C_3, and C_4 are adjusted such that C_1, C_2, C_3 control the slope of the transition line in the buoyancy-driven region (large values of X), while C_4 controls the slope of the transition line in inertia-driven region (small values of X).

The accuracy of the model of Bhagwat and Ghajar (2015a) was compared against the model of Taitel and Dukler (1976) for both horizontal and downward inclinations, while the model of Cheng et al. (1988), as recommended by the authors, was only used for the horizontal orientation. The comparisons were made using flow visualization data collected from more than 16 data sources consisting of 8 fluid combinations (air–water, air–oil, R-134a, R-113, R22/R-410A, air–K_2CO_3, air–glycerin, air–kerosene), pipe diameter in a range of 8.9 to 300 mm, liquid density in a range of 780–1420 kg/m^3, gas density in a range of 1.2–35 kg/m^3, liquid dynamic viscosity in a range of 0.0002–0.08 Pa·s, and all downward pipe inclinations including horizontal. For the details of these comparisons, refer to Bhagwat and Ghajar (2015a).

A graphical comparison between Taitel and Dukler (1976), Cheng et al. (1988), and Bhagwat and Ghajar (2015a) for horizontal and downward inclined two-phase flow of air–water in 12.7 mm I.D. pipe is shown in Figs. 3.15 and 3.16, respectively. The experimental data for stratified flow indicated by symbol "△" consists of both smooth and wavy stratified flows, while the non-stratified flow indicated by symbol "○" may consist of any or all of the bubbly, slug/intermittent, and annular flow patterns.

Note that Bhagwat and Ghajar (2015a) model given by Eq. (3.19) is of empirical form and merely attempts to mimic the trends of Taitel and Dukler (1976) model using $Fr_{SG} - X$ coordinates. Nevertheless, considering the practical difficulty of using a graphical/iterative solution involved in Taitel and Dukler (1976) model, Eq. (3.19) is explicit in nature and may be regarded as a quick method for hand calculations and to get an estimate of the existence of stratified flow in horizontal and downward pipe inclinations.

Transition Between Annular and Non-annular Flows Typically, annular flow shares the transition boundary with intermittent (churn/wavy annular) flow patterns; see Figs. 3.8, 3.9, 3.10, and 3.11. Moreover, in the case of horizontal and downward pipe inclinations, annular flow also shares a transition boundary with stratified flow

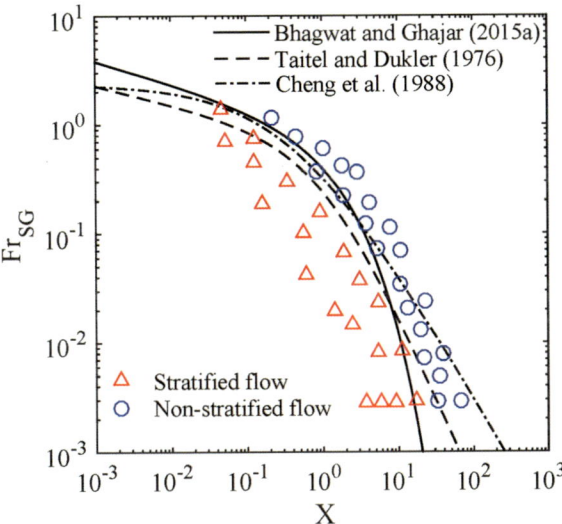

Fig. 3.15 Trends of different stratified flow transition models for horizontal two-phase flow of air–water in 12.7 mm I.D. pipe. (Based on data collected at the Two-Phase Flow Lab, OSU, Stillwater, OK)

pattern; see Figs. 3.10 and 3.11. In annular flow regime, the inertial forces greatly exceed the effect of gravitational forces, and hence it is usually assumed that the change in pipe orientation will have little effect on the transition between annular and other flow regimes. However, pipe diameter and fluid properties do affect the transition to annular flow regime, and hence, it is desired to have a transition equation for annular flow regime as a function of both pipe diameter and fluid properties. One such transition criterion proposed by Weisman and Kang (1981) is expressed by Eq. (3.24) and can be used for all pipe orientations. This criterion, however, at high liquid flow rates is found to incorrectly classify some region of the intermittent flow pattern as annular flow:

$$U_{SL} \geq U_{SG}\left[1.9\left(\frac{gD}{U_{SG}^2}\right)^{0.18}\left(\frac{g\Delta\rho\sigma}{U_{SG}\rho_G^2}\right)^{0.2}\right]^8 \qquad (3.24)$$

Another simple yet practical approach to predict the transition to annular flow regime is based on non-dimensional gas velocity (Froude number) corresponding to the point of pressure gradient minimum (concept discussed in Chap. 5). Two-phase flow literature for vertical upward flow reports that the pressure gradient minimum corresponds to the transition between churn/intermittent and annular flow regimes and the corresponding non-dimensional gas velocity satisfies Eq. (3.25). Based on the flow visualization and pressure drop measurements carried out at Two-Phase Flow Lab, OSU, it is found that the criteria given by Eq. (3.25) for transition to annular flow regime is a good approximation for the entire range of pipe orientations. The right- hand side of Eq. (3.25) gives a range between which the non-dimensional gas velocity can vary and for simplicity a value of 1.0 can be used. It also implies that

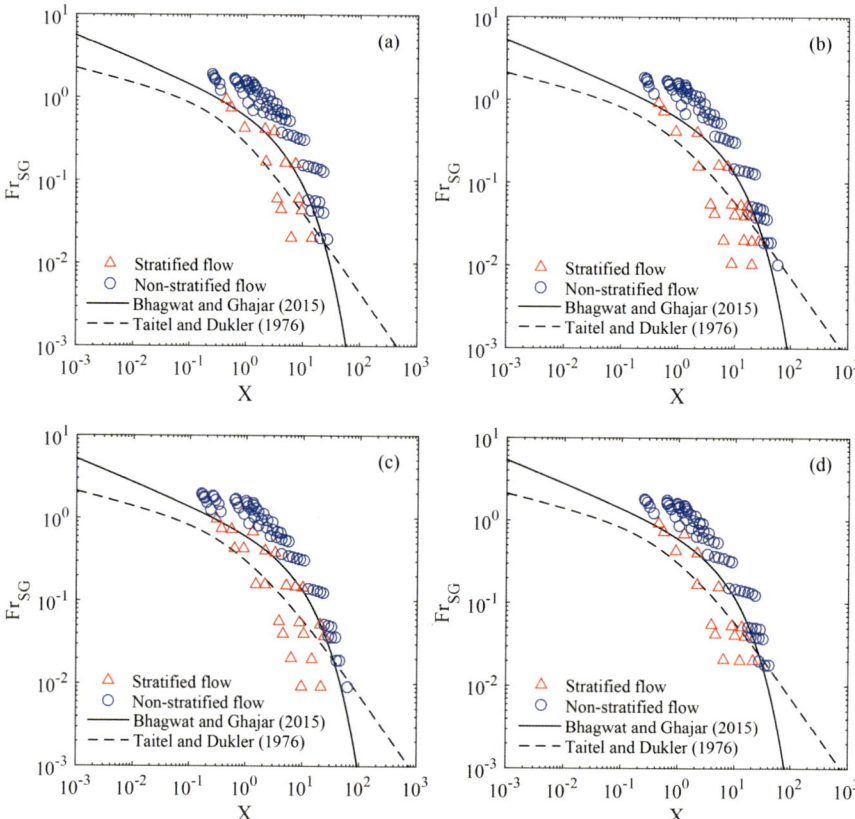

Fig. 3.16 Trends of different stratified flow transition models for downward inclined two-phase flow of air–water in 12.7 mm I.D. pipe, (**a**) $-10°$, (**b**) $-30°$, (**c**) $-45°$, (**d**) $-60°$. (Based on data collected at the Two-Phase Flow Lab, OSU, Stillwater, OK)

the transition to annular flow regime would be independent of liquid velocity. Based on experience, it is recommended that in conjunction with Eq. (3.25), additional simple criteria of $\alpha_G > 0.75$ can be used to confirm the existence of annular flow regime:

$$\frac{U_{SG}}{\sqrt{gD}}\sqrt{\frac{\rho_G}{(\rho_L - \rho_G)}} \geq 0.9 - 1.1 \qquad (3.25)$$

Taitel et al. (1980) present a criterion for transition from dispersed bubbly to annular flow regime. Their transition criteria based on force balance between the drag on liquid droplet and its weight lead to Eq. (3.26):

Table 3.1 Rule of thumb for qualitative classification of flow patterns

Void fraction (α_G)	Mass/Volumetric flow rate		Flow regime
	Gas phase	Liquid phase	
Low ($\alpha_G \leq 0.25$)	Low	Moderate/high	Bubbly
Low to moderate ($0 \leq \alpha_G \leq 0.25$)	Low	High	Dispersed bubbly
Moderate ($0.25 < \alpha_G \leq 0.75$)	Low	Low/moderate	Slug
Moderate ($0.25 < \alpha_G \leq 0.75$)	Moderate	Low/moderate	Intermittent
Moderate to high ($0.3 < \alpha_G \leq 0.9$)	Low/moderate	Low	Stratified
High ($0.75 < \alpha_G < 1$)	High	Low	Annular

$$\frac{U_{SG}\sqrt{\rho_G}}{(g\Delta\rho\sigma)^{0.25}} \geq 3.1 \qquad (3.26)$$

A quick comparison between Eqs. (3.25) and (3.26) shows that the superficial gas velocity (U_{SG}) predicted by Eq. (3.26) will always be higher than that predicted by Eq. (3.25). This is possibly because Eq. (3.25) gives transition to annular flow regime from churn/intermittent flow regimes at lower liquid flow rates, while Eq. (3.26) predicts transition from dispersed bubbly flow at high liquid flow rates and hence would need higher gas flow rates for transition to annular flow regime. In addition to the transition theories, Table 3.1 presents a rule of thumb based on the range of void fraction and qualitative range of gas and liquid flow rates associated with different flow patterns. It must be mentioned that this qualitative and quantitative range of two-phase flow variables is an approximate range based on experience and may give slight deviations compared to the real two-phase flow scenario.

Chapter 4
Void Fraction

Background

Void fraction (α_G) also referred to as liquid holdup ($\alpha_L = 1 - \alpha_G$) is defined based on its measurement technique such as local void fraction (using single-point probes), segmental void fraction (using gamma ray absorption method), cross-sectional void fraction (using capacitance probe), and volumetric void fraction (using quick closing valves). It must be noted that irrespective of the measurement technique, $\alpha_G + \alpha_L = 1$. Among these different methods, typically the cross-sectional ($\alpha_G = A_G/A$) and volumetric ($\alpha_G = V_G/V$) types of void fraction measurement are most preferred and of practical importance. Under the adiabatic two-phase flow condition over a short pipe length, it can be assumed that the two-phase flow structure remains unaltered throughout the pipe length, and cross-sectional void fraction can be equated to volumetric void fraction. The void fraction as a stand-alone physical parameter is typically of no use unless embedded in other constitutive equations to calculate parameters such as actual phase velocity, two-phase mixture density, and hence hydrostatic two-phase pressure drop and heat transfer, liquid height in stratified flow, and liquid film thickness in annular flow regime.

The void fraction in gas–liquid two-phase flow is found to be influenced by several parameters such as pipe orientation, flow rates of individual phases, fluid properties, and pipe diameter. These influences on the void fraction will be discussed next.

A. J. Ghajar, *Two-Phase Gas-Liquid Flow in Pipes with Different Orientations*,
SpringerBriefs in Applied Sciences and Technology,
https://doi.org/10.1007/978-3-030-41626-3_4

Effect of Pipe Orientation on Void Fraction

The effect of pipe orientation on the void fraction is essentially due to the change in flow patterns and influence of buoyancy force acting on the gas phase. Most of the experimental void fraction data available in the literature is concentrated over a narrow range of pipe orientations and phase flow rates (or alternatively the flow patterns). In this study, the void fraction was measured at 17 different pipe orientations ($0°$, $\pm5°$, $\pm10°$, $\pm20°$, $\pm30°$, $\pm45°$, $\pm60°$, $\pm75°$, and $\pm90°$) consisted of a wide range of void fraction ($0.09 \leq \alpha_G \leq 0.94$) and covered all important flow patterns observed in gas–liquid two-phase flow. The variation of void fraction with respect to change in pipe orientation is illustrated for different phase flow rates (Re_{SL} and Re_{SG}) in Figs. 4.1 and 4.2. The two liquid flow rates used in these figures are selected such that the representative data occupies the entire range of the void fraction and all flow patterns observed in this study.

Figure 4.1 shows the effect of pipe orientation on void fraction at low liquid flow rates ($Re_{SL} = 2000$) and a wide rage of gas flow rates ($Re_{SG} = 180 - 18,500$). As shown in the figure, the effect of pipe orientation on void fraction is significant (up to 100%) for low gas and liquid flow rates. These flow rates correspond to slug/ intermittent flow in upward pipe inclinations and stratified flow in downward pipe inclinations. At these flow rates, buoyancy acting on the gas phase aids the two-phase flow motion in upward pipe inclinations while resists its flow in downward

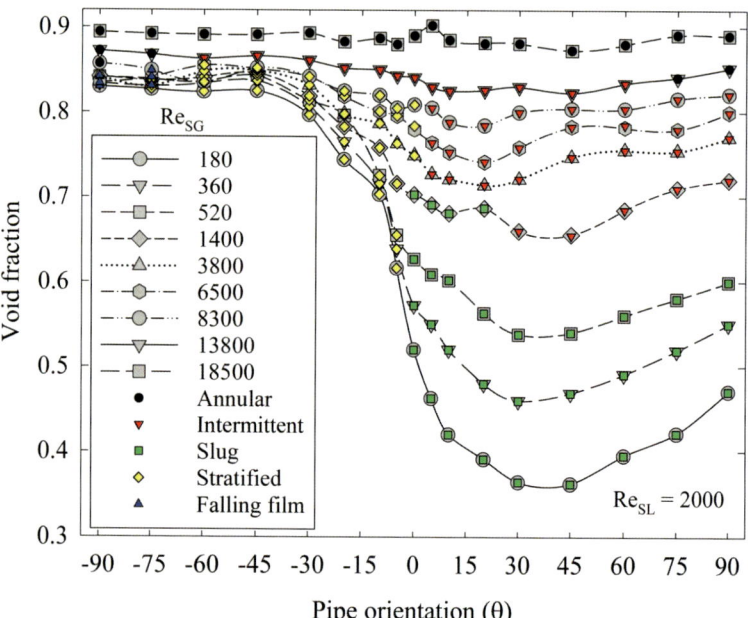

Fig. 4.1 Effect of pipe orientation on void fraction at low liquid flow rates. (Data from the Two-Phase Flow Lab, OSU, Stillwater, OK)

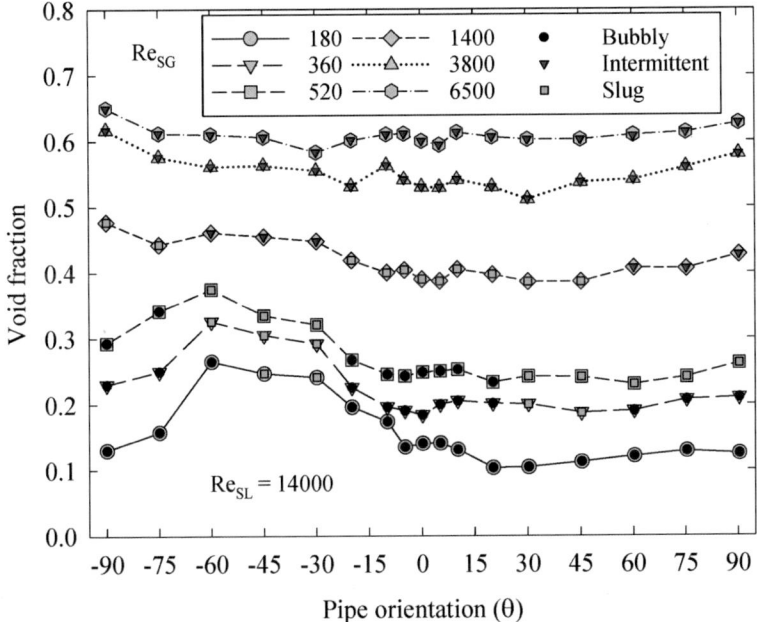

Fig. 4.2 Effect of pipe orientation on void fraction at high liquid flow rates. (Data from the Two-Phase Flow Lab, OSU, Stillwater, OK)

pipe inclinations increasing its residence time in the pipe and hence the void fraction. For a fixed flow pattern (slug/intermittent) in upward inclined flow, it is seen that the void fraction initially decreases up to $+30° < \theta < +45°$ and then increases again with increase in the pipe orientation. This trend of change in void fraction with change in pipe orientation is due to the similar trend of the translational velocity of the gas slug (as a function of pipe orientation) that decides the residence time of the gas phase in the pipe and hence the void fraction. The effect of pipe orientation on void fraction is found to vanish for the inertia-driven region of intermittent (wavy annular) and annular two-phase flow. Moreover, the void fraction in stratified flow is apparently insensitive to the change in pipe orientation. At moderate gas and liquid flow rates, void fraction in pipe orientations steeper than $-45°$ drops down due to the change in flow pattern from stratified to slug/intermittent at similar phase flow rates.

Figure 4.2 shows the effect of pipe orientation on void fraction at high liquid flow rates ($Re_{SL} = 14,000$) and a wide rage of gas flow rates ($Re_{SG} = 180 - 6500$). As shown in the figure, at high liquid flow rates, the effect of pipe orientation on void fraction gradually fades away; however a similar type of flip in void fraction trend as mentioned above occurs after $-60°$ of pipe orientations. This change in trend of void fraction is again due to dissimilar flow patterns that may exist for similar phase flow rates but different pipe orientations. It is also clear from these trends that the void fraction in downward pipe inclinations is always greater than that in upward pipe

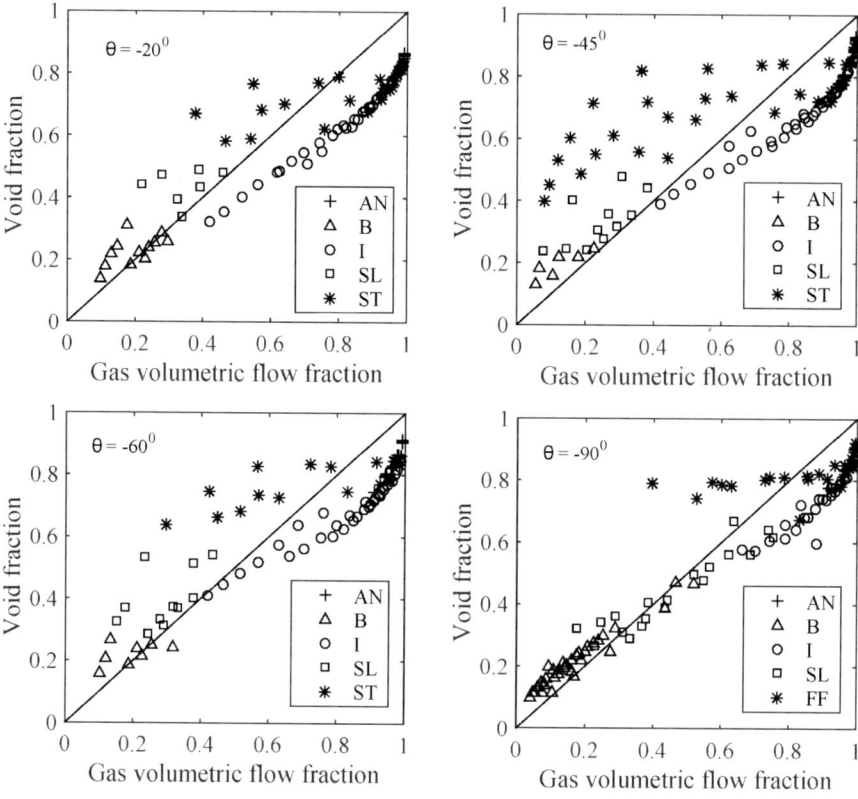

Fig. 4.3 Void fraction and gas volumetric flow fraction relationship in downward pipe inclinations (AN annular, B bubbly, FF falling film, I intermittent, SL slug, ST stratified). (Data from the Two-Phase Flow Lab, OSU, Stillwater, OK)

inclinations essentially due to the resistance offered by the buoyancy force to the gas phase.

Considering the overall effect of the pipe orientation on the void fraction, three different relationships between α_G and λ can be deduced such that for buoyancy-driven flows $\alpha_G > \lambda$ and for inertia-driven flows $\alpha_G < \lambda$ hold true. The third relationship would be $\alpha_G = \lambda$ when the change in buoyancy to inertia-driven flow takes place. The relationship $\alpha_G > \lambda$ implies that for buoyancy-driven two-phase flow in downward pipe inclinations, the gas phase travels at a lower velocity than the liquid phase and hence results into higher void fraction values. Note that for all upward pipe inclinations, gas phase moves faster than the liquid phase giving a positive slippage at the gas–liquid interface, and hence $\alpha_G < \lambda$ holds true. These three different cases of $\alpha_G - \lambda$ relationship that may exist in downward pipe inclinations are illustrated in Fig. 4.3. In general, the flow patterns/pipe orientations that may correspond to these different cases of $\alpha_G - \lambda$ are summarized in Table 4.1.

Table 4.1 Summary of two-phase flow situations corresponding to $\alpha_G - \lambda$ relationships

Condition	Flow patterns/Pipe orientation
$\alpha_G < \lambda (U_G > U_L, \; S > 1)$	All flow patterns in upward pipe inclination
$\alpha_G = \lambda (U_G \approx U_L, \; S \approx 1)$	Homogeneous flow (dispersed bubbly, annular mist)
$\alpha_G > \lambda (U_G < U_L, \; S < 1)$	Stratified, slug flow $Fr_{SG} < 0.1$ and $\theta < 0°$

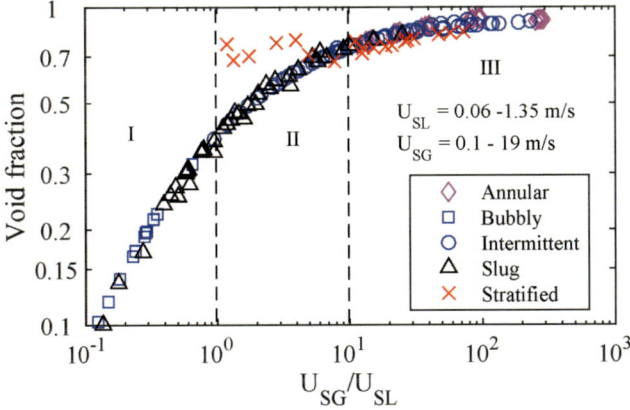

Fig. 4.4 Variation of void fraction with change in gas and liquid flow rates. (Data from the Two-Phase Flow Lab, OSU, Stillwater, OK)

Effect of Phase Flow Rates on Void Fraction

The effect of phase flow rate on void fraction is indirectly through the change in flow pattern with change in phase flow rates. Figure 4.4 shows the effect of change in gas and liquid flow rate on void fraction. The void fraction data is plotted against a non-dimensional phase velocity ratio (U_{SG}/U_{SL}) to reduce the scatter of the data. Broadly, the void fraction as a function of phase flow rates can be divided into three regions shown in Fig. 4.4. In region I occupied by bubbly and slug flow patterns, for a fixed liquid flow rate, void fraction increases rapidly with a small increase in the gas flow rate, whereas in region III occupied by intermittent, stratified, and annular flow patterns, void fraction remains virtually independent of change in gas- and liquid-phase flow rates. Region II is a mixed bag of slug and intermittent and stratified flow patterns, and its slope merges smoothly with those in regions I and III. Compared to other flow patterns, the void fraction corresponding to stratified flow shows a different behavior in region II. Void fraction in stratified flow regime remains relatively insensitive to the change in phase flow rates for both regions II and III. Similar conclusions could be drawn by plotting the void fraction against two-phase flow quality, a well-adopted presentation style in high system pressure two-phase flows.

The variation of void fraction with change in two-phase flow quality is depicted in Fig. 4.5 for (a) constant system pressure of 1 MPa and varying slip ratios ($S = 1$ to

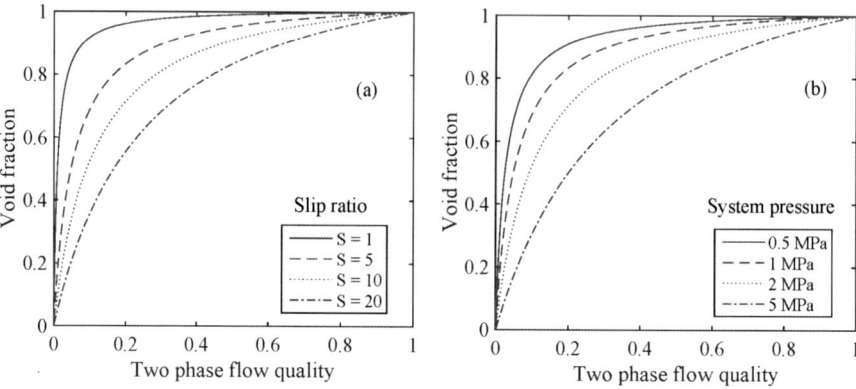

Fig. 4.5 Variation of void fraction with change in two-phase flow quality. (**a**) System pressure = 1 MPa. (**b**) Slip ratio = 2. (Data from the Two-Phase Flow Lab, OSU, Stillwater, OK)

20) and (b) constant slip ratio ($S = 2$) and varying system pressure (0.5 MPa to 5 MPa). For fixed two-phase flow conditions such as system pressure and quality, an increase in slip ratio reduces the void fraction; see Fig. 4.5a. For these fixed conditions, the maximum void fraction is essentially the homogeneous void fraction obtained for no-slip condition ($S = 1$). In the case of fixed slip ratio, an increase in system pressure decreases the specific volume occupied by the gas phase (due to increase in gas-phase density) and hence reduces the void fraction; see Fig. 4.5b.

Effect of Fluid Properties on Void Fraction

The two important fluid properties that are known to affect the void fraction are the gas-phase density and the liquid-phase dynamic viscosity. The effect of gas-phase density on the void fraction is essentially in a case of high system pressure flow such as that observed in nuclear cooling applications and the refrigeration industry, whereas the effect of liquid-phase dynamic viscosity on the void fraction is of significant interest in chemical engineering processes and the oil and gas industry applications that involve viscous oils having dynamic viscosity up to an order of 500 compared to that of liquid water at room temperature and pressure. The increase in system pressure or alternatively increase in the gas-phase density reduces the slippage between the two phases and also reduces the void fraction measured at similar mass flow rate as was shown in Fig. 4.5b. A denser gas phase (lower specific volume) will occupy less volume in the test section and hence would result in reduced void fraction. As shown in Fig. 4.6, the slope of void fraction vs. quality is a function of the phase density ratio, and in the limiting case of $\rho_G \approx \rho_L$, there is a negligible slippage between the two phases, and void fraction varies linearly with change in two-phase flow quality ($\alpha_G = x$). In this case, the two-phase flow void fraction becomes equal to the homogeneous flow void fraction. A similar trend of

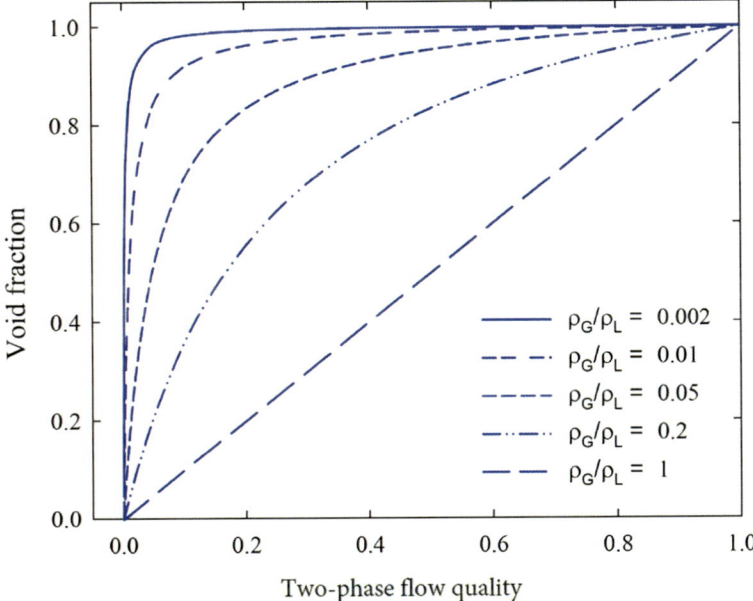

Fig. 4.6 Variation of void fraction with respect to change in quality as a function of phase density ratio. (Data from the Two-Phase Flow Lab, OSU, Stillwater, OK)

variation of void fraction as a function of gas-phase density for two-phase flow of refrigerants has been reported by Shedd (2010) and Keinath (2012). They found the void fraction to decrease with increase in refrigerant saturation temperature (increase in saturation or system pressure).

In addition to the gas-phase density, void fraction is also known to be affected by the change in liquid dynamic viscosity. Gokcal (2008) and Jeyachandra (2011) measured the void fraction in a 50.2 mm I.D. pipe using air and oil of dynamic viscosity in a range of 0.18 Pa-s to 0.58 Pa-s. Overall they found the void fraction to slightly decrease with increase in the liquid dynamic viscosity. Similar results are reported by Oshinowo (1971) who measured void fraction in vertical upward and downward pipes using air–water and air and four different concentrations of ethylene glycol. Mukherjee (1979) measured void fraction using air–kerosene and air–oil fluid combinations in upward and downward pipe inclinations. The results of these experiments revealed that the void fraction is inversely proportional to the liquid dynamic viscosity as shown in Fig. 4.7. From the results of these studies, it was concluded that the increase in liquid viscosity increases the viscous drag on liquid phase and reduces its velocity causing it to accumulate in the pipe and hence results in decrease in void fraction due to increase in liquid holdup.

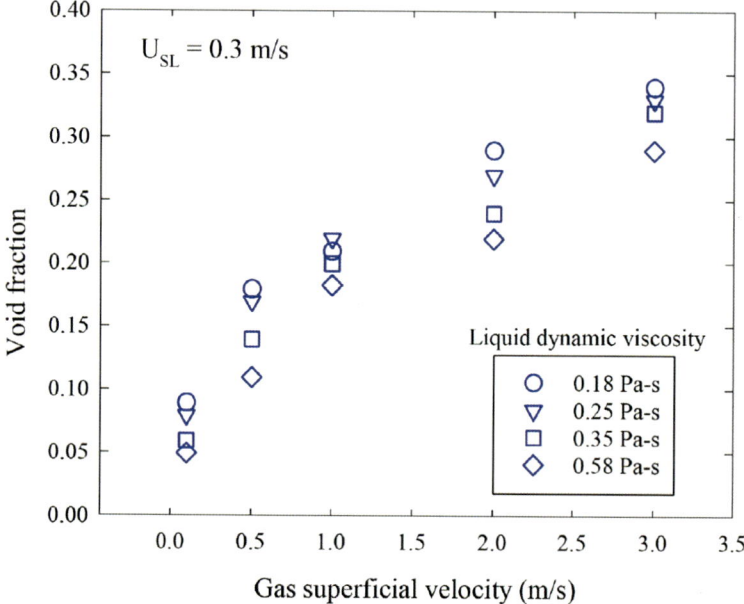

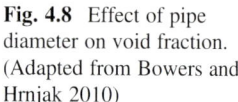

Fig. 4.7 Effect of liquid dynamic viscosity on void fraction. (Adapted from Gokcal 2008)

Fig. 4.8 Effect of pipe
diameter on void fraction.
(Adapted from Bowers and
Hrnjak 2010)

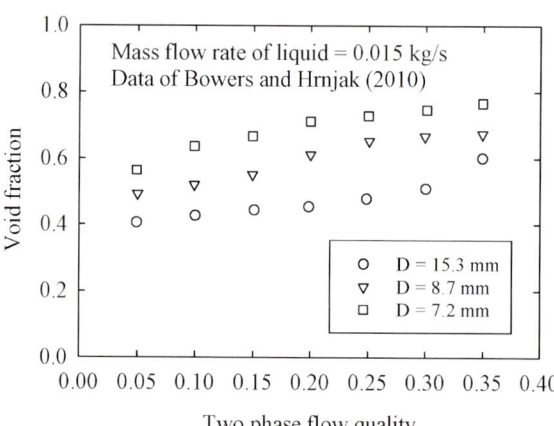

Two phase flow quality

Effect of Pipe Diameter on Void Fraction

Effect of pipe diameter on void fraction is reflected in the form of pipe wall-induced
drag exerted on the gas phase. In comparison to large pipes, small diameter pipes
offer more resistance to the motion of gas phase increasing its residence time and
hence the local and averaged void fraction in pipes. Experimental data of Bowers
and Hrnjak (2010) in Fig. 4.8 show about 20–80% increase in measured void

fraction for a decrease in pipe diameter from 15.3 to 7.2 mm. The void fraction as a function of pipe diameter also depends upon phase flow rates (flow patterns/mass flux/quality) such that the effect of change in pipe diameter on void fraction is most prominent for bubbly/slug/intermittent flows, whereas, for the high values of void fraction corresponding to that in annular flow regime (shear driven flow) typically $\alpha_G > 0.8$, the void fraction is relatively insensitive to the pipe diameter. This observation is in agreement with the work of Kaji and Azzopardi (2010) who investigated the effect of pipe diameter in a range of 5–70 mm on void fraction in vertical upward annular flow regime for non-boiling two-phase flow for similar gas and liquid superficial velocities. They found that the effect of pipe diameter on void fraction in the annular flow regime is negligible. The higher values of void fraction in small diameter pipes can be justified considering the flow structure and wall effects on the gas phase. In the case of two-phase flow through small diameter pipes, typically at low gas flow rates, the two-phase flow is in form of bubble train (quasi-spherical or equal length elongated bubbles that occupy the entire pipe cross section), whereas the flow structure in large diameter pipes may be in the form of dispersed bubbles. This relationship between void fraction and pipe diameter may not be applicable to microscale two-phase flow.

The parametric analysis of void fraction reveals that the void fraction must be modeled to consider the effect of flow patterns, pipe orientation, pipe diameters, and fluid properties. What follows is a brief review of some of the void fraction modeling methods available in the two-phase flow literature and recommendation of well-scrutinized void fraction correlations applicable for a wide range of two-phase flow conditions.

Modeling of Void Fraction

Two-phase flow literature reports plethora of void fraction models/correlations; however, most of these models are confined to certain flow patterns, pipe orientation, and fluid properties. Based on the general physical form of these correlations, they can be broadly classified as those based on separated flow models, drift flux models, and empirical models. A comprehensive review of these different types of models is presented by Woldesemayat and Ghajar (2007), Godbole et al. (2011), and Ghajar and Bhagwat (2014b). Among all these models, the drift flux model was identified to be more versatile, accurate, and flexible. It was also one of the most preferred models for several industrial applications. For these reasons, the drift flux model was used in our studies. Following is a brief summary of the physical structure of the drift flux model-based void fraction correlation.

Drift Flux Model (DFM) The general concept of the DFM was first envisaged by Zuber and Findlay (1965) and later developed and contributed by Wallis (1969) and Ishii (1977). The DFM assumes one phase dispersed in other continuous phase and requires the determination of distribution parameter (C_o) and drift velocity (U_{GM}) as variables to calculate the void fraction. The DFM is best applicable for one-dimensional flows and is usually not recommended for flow patterns involving significant interfacial slippage such as stratified and annular flow patterns. The flow patterns such as bubbly flow, slug flow, and mist flow are the preferred flow patterns to be modeled using the concept of drift flux. However, with appropriate formulations that consider the two-phase flow physics, the application of DFM can also be extended to other flow patterns such as annular flow. The general form of DFM to calculate void fraction is presented by Eq. (4.1) where $U_M = U_{SL} + U_{SG}$:

$$\langle \alpha_G \rangle = \frac{\langle U_{SG} \rangle}{C_o \langle U_M \rangle + \langle \langle U_{GM} \rangle \rangle} \tag{4.1}$$

The terms involving $\langle \rangle$ represent cross-sectional averaged properties, while $\langle \langle \rangle \rangle$ indicate cross-sectional and void fraction weighted averaged properties. The distribution parameter C_o is a representation to account for the distribution of the gas phase across the pipe cross section (concentration profile). It also serves as a correction factor to the homogeneous flow theory (which assumes no local slip between the two phases) to acknowledge the fact that the concentration profile and the two-phase flow velocity profile can vary independently of each other across the pipe cross section, whereas the physical interpretation of drift velocity (U_{GM}) is the cross-sectional void fraction weighted average of the local relative velocity of the gas phase with respect to the two-phase mixture velocity at the pipe volume center. The local relative motion between the gas- and the two-phase mixtures is considerable and uniform across the pipe cross section when there is a strong coupling between the two phases as in the case of the flow of dispersed bubbles in continuous liquid medium, that is, dispersed bubbly flow. As the two-phase flow transits to annular flow regime, the local relative velocity of the gas phase with respect to the two-phase mixture at the pipe volume center becomes negligible and so does the drift velocity. Although the one-dimensional DFM is used in context of cross-sectional averaged void fraction, there are several DFM-based correlations available in the literature that can be practically implemented for void fraction calculations based on volumetric void fraction experimental data instead of the cross-sectional void fraction. The parity between the cross-sectional and volumetric void fraction holds true in the case of non-boiling two-phase flow (two-component two-phase flow) where the cross-sectional distribution of the gas phase with respect to the liquid phase remains virtually unaltered over a short length of pipe. Henceforth, without making any distinction between the cross-sectional and volumetric void fraction, the void fraction is simply expressed as $\langle \alpha_G \rangle = \alpha_G$. Similar justification is applicable for all the cross-sectional averaged quantities involved in Eq. (4.1). The physical form of Eq. (4.1) resembles the equation of a straight line ($y = mx + c$) shown in Fig. 4.9 such that the distribution parameter C_o is analogous to the slope "m" and the drift velocity is analogous to the y-intercept "c." Thus, the distribution parameter and the drift velocity could be obtained by plotting the data with $U_G = U_{SG}/\alpha_G$ and U_M coordinates and then finding the slope and intercept of the best-fitting line. However, this

Fig. 4.9 Graphical
representation of drift flux
model

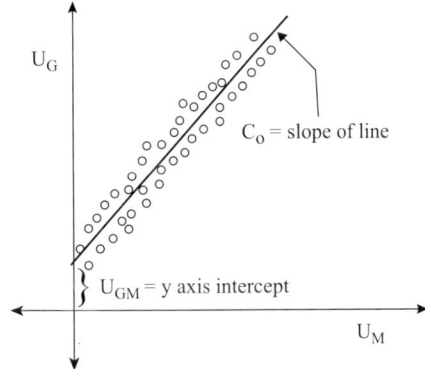

method is highly impractical since the slope and intercept of line would depend on
the scatter of data and cannot be used in case of real-time applications where void
fraction information is not known beforehand. Obviously, another approach to
determine C_o and U_{GM} is to use empirical correlations (as a function of two-phase
flow variables) available in the two-phase flow literature. What follows next is a brief
presentation of some of the well-scrutinized and validated flow pattern-specific and
flow pattern-independent correlations to determine the distribution parameter C_o and
the drift velocity U_{GM} used in the general form of DFM, Eq. (4.1), to calculate void
fraction.

Hibiki and Ishii (2003) Correlation for Vertical Upward Bubbly Flow The
distribution parameter and drift velocity equations given by Hibiki and Ishii
(2003) for vertical upward bubbly flow are given in Eq. (4.2). To use this equation,
the existence of bubbly flow could be determined by first using flow pattern
transition equation given by Eq. (3.4).

$$\left.\begin{array}{l} C_o = \left(1.2 - 0.2\sqrt{\dfrac{\rho_G}{\rho_L}}\right)(1 - \exp(-18\alpha_G)) \\[4mm] U_{GM} = \sqrt{2}\left(\dfrac{\sigma g(\rho_L - \rho_G)}{\rho_L^2}\right)^{0.25}(1 - \alpha_G)^{1.75} \end{array}\right\} \begin{array}{l} \text{(bubbly flow)} \\[2mm] \theta = +90^\circ \end{array} \qquad (4.2)$$

Gomez et al. (2000) Correlation for Bubbly Flow Another most referred void
fraction model for bubbly flow is that of Gomez et al. (2000) given by Eq. (4.3). This
model is essentially developed for upward pipe inclinations, and it may not repro-
duce correct results if used for horizontal two-phase flow since Eq. (4.3) would give
$U_{GM} = 0$. Moreover, Gomez et al. (2000) have recommended use of a constant to
represent distribution parameter that does not account for the effect of fluid properties, flow properties, and pipe orientation on the distribution parameter. Although
their void fraction model is developed for bubbly flow, our experience shows that
this correlation also works well for the slug flow regime. Note that to use Eq. (4.3),

the existence of bubbly flow can be confirmed by using bubbly-slug flow transition model given by Eq. (3.1):

$$
\left.\begin{array}{l}
C_{\mathrm{o}} = 1.15 \\[2mm]
U_{\mathrm{GM}} = 1.53\left(\frac{g\sigma(\rho_{\mathrm{L}}-\rho_{\mathrm{G}})}{\rho_{\mathrm{L}}^2}\right)^{0.25}\sqrt{1-\alpha_{\mathrm{G}}}\sin\theta
\end{array}\right\}
\begin{array}{l}
(\text{bubbly, slug flow}) \\[2mm]
0^{\circ} < \theta \le +90^{\circ}
\end{array}
\qquad (4.3)
$$

Hibiki and Ishii (2003) Correlation for Slug Flow For slug flow regime, Hibiki and Ishii (2003) proposed expressions for distribution parameter and drift velocity given by Eq. (4.4). The drift velocity used in their equations is essentially the modified form of proportionality (between pipe diameter and slug translational velocity) proposed by Nicklin et al. (1962). These correlations given by Eqs. (4.2) and (4.4) are valid for small diameter pipes ($10 < D_{\mathrm{h}} < 50$ mm). For large diameter pipes, Kataoka and Ishii (1987) have proposed sets of equations given by Eq. (4.5) to account for the effect of pipe diameter and fluid properties on the drift velocity. The non-dimensional hydraulic pipe diameter (D_{h}^{+}), non-dimensional drift velocity (U_{GM}^{+}), and viscosity number (N_{μ}) are defined by Eqs. (4.6), (4.7), and (4.8), respectively. Their equation is developed based on data of air–water and steam–water vertical upward two-phase flow consisting of data with $20 < D_{\mathrm{h}} < 240$ mm and $0.1 \le p_{\mathrm{SYS}} \le 18$ MPa. The distribution parameter of Kataoka and Ishii (1987) correlation is the same as that of Eq. (4.4). In the case of two-phase flow through rectangular ducts, the constants "1.2" and "0.2" in Eq. (4.4) are replaced by "1.35" and "0.35", respectively:

$$
\left.\begin{array}{l}
C_{\mathrm{o}} = \left(1.2 - 0.2\sqrt{\dfrac{\rho_{\mathrm{G}}}{\rho_{\mathrm{L}}}}\right) \\[4mm]
U_{\mathrm{GM}} = 0.35\sqrt{\dfrac{gD(\rho_{\mathrm{L}}-\rho_{\mathrm{G}})}{\rho_{\mathrm{L}}}}
\end{array}\right\}
\begin{array}{l}
(\text{slug flow}) \\[3mm]
\theta = +90^{\circ}
\end{array}
\qquad (4.4)
$$

$$
U_{\mathrm{GM}}^{+} =
\begin{cases}
0.00019\left(D_{\mathrm{h}}^{+}\right)^{0.809}\left(\dfrac{\rho_{\mathrm{G}}}{\rho_{\mathrm{L}}}\right)^{-0.157} N_{\mu}^{-0.562} & : N_{\mu} \le 2.25 \times 10^{-3},\ D_{h}^{+} > 30 \\[4mm]
0.030\left(\dfrac{\rho_{\mathrm{G}}}{\rho_{\mathrm{L}}}\right)^{-0.157} N_{\mu}^{-0.562} & : N_{\mu} \le 2.25 \times 10^{-3},\ D_{h}^{+} \le 30 \\[4mm]
0.92\left(\dfrac{\rho_{\mathrm{G}}}{\rho_{\mathrm{L}}}\right)^{-0.157} & : N_{\mu} > 2.25 \times 10^{-3}
\end{cases}
$$

$$
\qquad (4.5)
$$

$$
D_{\mathrm{h}}^{+} = \frac{D_{\mathrm{h}}}{\sqrt{\sigma/(g\Delta\rho)}}
\qquad (4.6)
$$

$$U_{GM}^{+} = \frac{U_{GM}}{((\sigma g \Delta \rho)/\rho_L^2)^{0.25}} \tag{4.7}$$

$$N_\mu = \frac{\mu_L}{\left(\rho_L \sigma \sqrt{\sigma/(g\Delta\rho)}\right)^{0.5}} \tag{4.8}$$

Rouhani and Axelsson (1970) Correlation The DFM-based void fraction correlation of Rouhani and Axelsson (1970) given by Eqs. (4.9) and (4.10) is one of the most preferred correlations used in refrigeration industry to estimate void fraction in evaporators and condensers. The term $(1 - x)$ ensures decrease of distribution parameter with increase in two-phase flow quality or alternatively shift of flow pattern from bubbly/slug to annular flow:

$$C_o = \begin{cases} 1 + 0.2(1-x)\left(\frac{gD\rho_L^2}{G^2}\right)^{0.25} & : \theta = +90°, \ 0.1 \leq p_{sys} \leq 14 \ \text{MPa} \\ 1 + 0.12(1-x) & : \theta = 0°, \ 0.1 \leq p_{sys} \leq 14 \ \text{MPa} \end{cases} \tag{4.9}$$

$$U_{GM} = 1.18\left[\frac{g\sigma(\rho_L - \rho_G)}{\rho_L^2}\right]^{0.25} \tag{4.10}$$

Woldesemayat and Ghajar (2007) Correlation To get rid of the flow pattern dependency, Woldesemayat and Ghajar (2007) proposed equations for distribution parameter and drift velocity as a function of fluid properties, pipe diameter, and pipe orientation shown in Eqs. (4.11) and (4.12). Their correlation is a modification of Dix (1971) correlation for two-phase flow over rod bundles and is independent of flow patterns. Correlation of Woldesemayat and Ghajar (2007) is verified against 2845 data points from 8 different sources for horizontal (900 data points), upward inclined (1542 data points), and upward vertical (403 data points) pipe inclinations. The data covered pipe diameters in a range of 12–80 mm and different fluid combinations (natural gas–water, air–water, air–kerosene). Figure 4.10 shows the comparison of their correlation with the entire set of experimental data for horizontal, upward inclined, and upward vertical pipe inclinations. Correlation of Woldesemayat and Ghajar (2007) predicted 79% and 86% of the entire data set within an accuracy of ±10% and ±15%, respectively. In the case of unknown flow pattern in horizontal and upward pipe inclinations, Woldesemayat and Ghajar (2007) correlation can be used with a reasonable accuracy (except for stratified flow). Note that in Eq. (4.12), the multiplying factor of 2.9 has units of $m^{-0.25}$:

$$C_o = \lambda\left(1 + \left[\frac{U_{SL}}{U_{SG}}\right]^{(\rho_G/\rho_L)^{0.1}}\right) \tag{4.11}$$

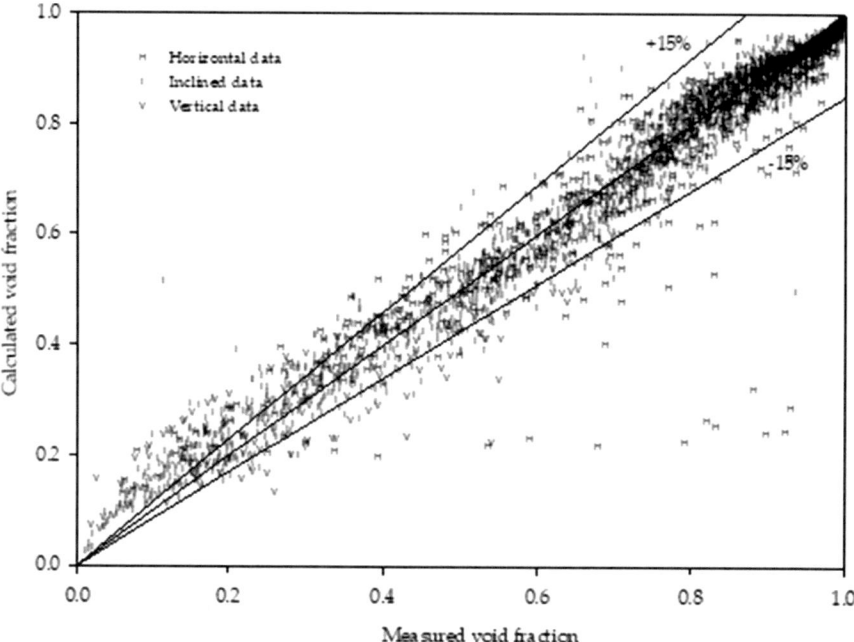

Fig. 4.10 Comparison of Woldesemayat and Ghajar (2007) void fraction correlation with 2845 experimental data points for horizontal, upward inclined, and upward vertical pipe inclinations

$$U_{GM} = 2.9 \left[\frac{gD\sigma(1 + \cos\theta)(\rho_L - \rho_G)}{\rho_L^2} \right]^{0.25} (1.22 + 1.22\sin\theta)^{P_{atm}/P_{sys}} \qquad (4.12)$$

Bhagwat and Ghajar (2014) Correlation A fairly recent void fraction correlation developed by Bhagwat and Ghajar (2014) considers the effect of flow patterns (in form of phase flow rates), pipe orientation, pipe diameter, and fluid properties on the distribution parameter (C_o) and drift velocity (U_{GM}) as shown in Eqs. (4.13) and (4.16), respectively. The experimental data base used for performance verification of their void fraction correlation consists of 8255 data points, with 4277 data points for air–water, natural gas–water, and argon–water fluid combinations, 1110 data points for liquid refrigerants and their vapors (R11, R12, R22, R134a, R114, R410A, R290, and R1234yf), 570 data points for steam–water, 567 data points for air–oil with liquid-phase dynamic viscosity ranging from 0.005 Pa-s to 0.6 Pa-s, and 1731 data points for other fluid combinations such as air–kerosene, air–glycerin, argon–ethanol, argon–acetone, and argon–alcohol. The experimental data listed above belongs to more than 60 sources in the two-phase flow literature and consists of pipe orientations in a range of $-90°$ to $+90°$ and hydraulic pipe diameters from 0.5 mm to 305 mm covering the micro- to macroscale two-phase flow phenomenon. The experimental void fraction data for refrigerants and steam–water consists of both adiabatic and diabatic (boiling and condensing) flow conditions. In fact, about 36%

Table 4.2 Application range of two-phase flow parameters for Bhagwat and Ghajar (2014) void fraction correlation

Parameter	Range
Pipe diameter (mm)	0.5–305
Pipe orientation (θ)	$-90° \leq \theta \leq +90°$
Density ratio (ρ_L/ρ_G)	6–875
Liquid dynamic viscosity (Pa·s)	0.0005–0.6
Mixture mass flux (kg/m²·s)	10–8450
Two-phase quality (x)	0–1
Two-phase Reynolds number (Re_M)	$10-5 \times 10^6$
Pipe geometry	Circular, rectangular, annular
Flow patterns	All (except stratified flow with $Fr_{SG} \leq 0.1$ in $\theta < 0°$)

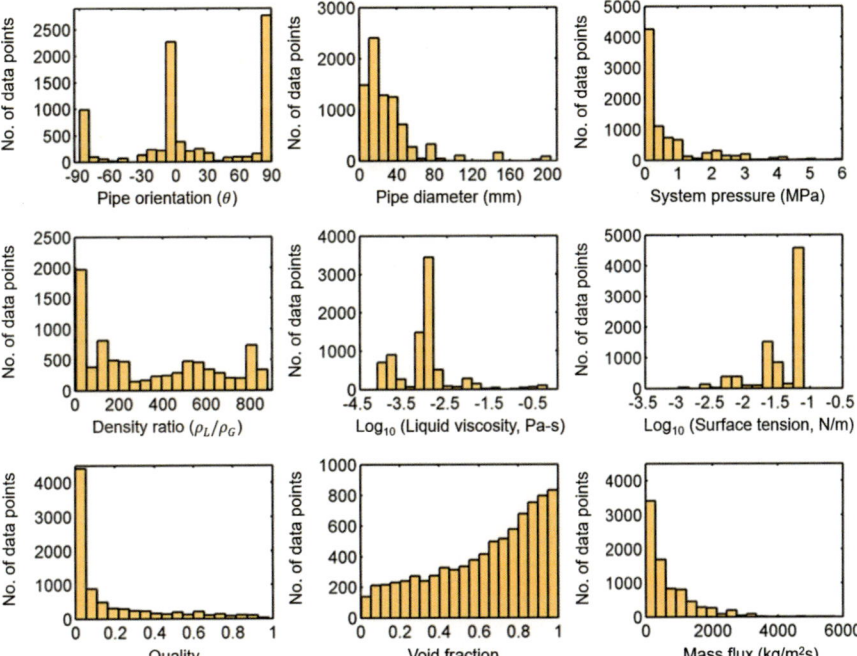

Fig. 4.11 Distribution of the experimental data used in the development of Bhagwat and Ghajar (2014) void fraction correlation

of the refrigerant data and 62% of the steam–water data are for diabatic conditions. Moreover, the diabatic two-phase flow data for steam–water consists of noncircular (rectangular) pipes with dimensions 11.1 mm × 93.6 mm, 6.35 mm × 50.8 mm, and 12.7 mm × 50.8 mm. Their void fraction correlation is based on the most comprehensive data bank and is the most robust correlation applicable for a wide range of two-phase flow conditions. Table 4.2 provides the application range of their correlation, and Fig. 4.11 depicts the distribution of the experimental data over different two-phase flow parameters. Since the experimental data for very large pipe diameter ($D_h = 305$ mm) and very high system pressure ($p_{sys} \approx 18$ MPa) is limited compared to other experimental data, they are not included in the histograms shown in Fig. 4.11.

The variable $C_{o,1}$ used in Eq. (4.13) for the distribution parameter (C_o) is defined by Eq. (4.14) and is modeled as a variable such that it may vary between 0.2 and 0 to ensure the variation of distribution parameter for bubbly flow $C_o \approx 1.2$ to annular flow where $C_o \to 1.0$. More details on the development of Eqs. (4.13) and (4.14) may be obtained from Bhagwat and Ghajar (2014):

$$C_o = \frac{2 - (\rho_G/\rho_L)^2}{1 + (Re_M/1000)^2}$$

$$+ \frac{\left[\sqrt{(1 + (\rho_G/\rho_L)^2 \cos\theta)/(1 + \cos\theta)}^{(1-\alpha_G)} \right]^{2/5} + C_{o,1}}{1 + (1000/Re_M)^2} \tag{4.13}$$

where two-phase mixture Reynolds number (Re_M) is given by Eq. (4.15).

$$C_{o,1} = \left(c_1 - c_1 \sqrt{\rho_G/\rho_L} \right) \left[(2.6 - \lambda)^{0.15} - \sqrt{f_M} \right] (1 - x)^{1.5} \tag{4.14}$$

The constant c_1 is 0.2 for circular pipes and 0.4 for rectangular and annular ducts. Note that the two-phase friction factor (f_M) in Eq. (4.14) is based on two-phase mixture Reynolds number (Re_M) given by Eq. (4.15) and could be calculated using appropriate single-phase friction factor equation for laminar and turbulent flows:

$$Re_M = \frac{(U_{SL} + U_{SG})\rho_L D_h}{\mu_L} \tag{4.15}$$

The expression for drift velocity (U_{GM}) used by Bhagwat and Ghajar (2014) as given in Eq. (4.16) is essentially based on the translational velocity of the gas slug in different pipe inclinations suggested by Bendiksen (1984). However, instead of using the constant of 0.54 as originally proposed by Bendiksen (1984), a proportionality constant of 0.45 is used since it gave a better accuracy in overall prediction of the void fraction. The correction factors c_2 and c_3 used in Eq. (4.16) account for the effect of increase in liquid viscosity and pipe diameter on the reduction in drift velocity. These correction factors are expressed by Eqs. (4.17) and (4.18). More details and justification about the effect of these parameters on drift velocity and hence the use of these correction factors could be obtained from Gokcal et al. (2009), Kataoka and Ishii (1987), and Bhagwat and Ghajar (2014). The pipe diameter beyond which the correction to drift velocity in form of c_3 is deemed necessary is identified using Laplace number (non-dimensional pipe diameter) defined by Eq. (4.19):

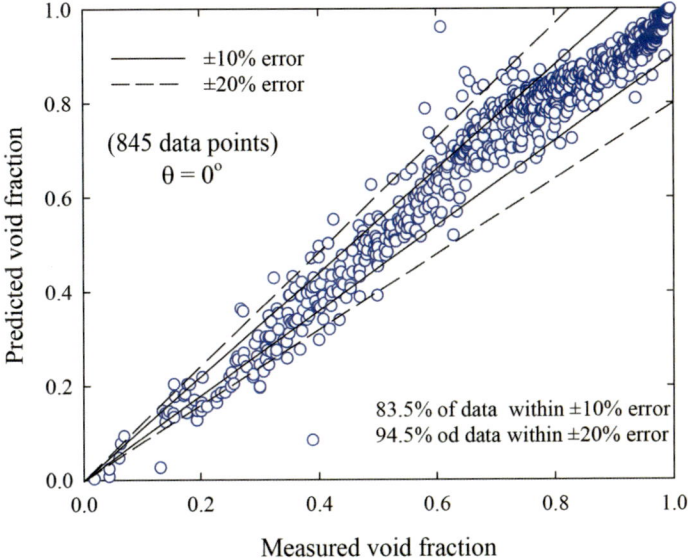

Fig. 4.12 Performance of the Bhagwat and Ghajar (2014) void fraction correlation against air–water data for horizontal pipe inclination. (Data from the Two-Phase Flow Lab, OSU, Stillwater, OK)

$$U_{GM} = (0.35 \sin\theta + 0.45 \cos\theta)\sqrt{\frac{gD_h(\rho_L - \rho_G)}{\rho_L}}(1 - \alpha_G)^{0.5}c_2 c_3 \qquad (4.16)$$

$$c_2 = \begin{pmatrix} \left(\dfrac{0.434}{\log_{10}(\mu_L/0.001)}\right)^{0.15} & : (\mu_L/0.001) > 10 \\ 1 & : (\mu_L/0.001) \leq 10 \end{pmatrix} \qquad (4.17)$$

$$c_3 = \begin{pmatrix} (La/0.025)^{0.9} & : La < 0.025 \\ 1 & : La \geq 0.025 \end{pmatrix} \qquad (4.18)$$

$$La = \sqrt{\frac{\sigma}{g(\rho_L - \rho_G)}}\frac{1}{D_h} \qquad (4.19)$$

Graphical performance of void fraction correlation of Bhagwat and Ghajar (2014) for OSU's air–water data in horizontal, vertical upward, and vertical downward pipe orientation is shown in Figs. 4.12 and 4.13. As shown in Fig. 4.12, for horizontal flow, their correlation predicts more than 90% of data points within ±20% error bands, while for vertical upward and downward flow (see Fig. 4.13), their correlation predicts 88.3% and 94.6% of data within ±20% error bands, respectively.

The graphical performance of Bhagwat and Ghajar (2014) void fraction correlation for air–water void fraction data in upward and downward inclined flow is

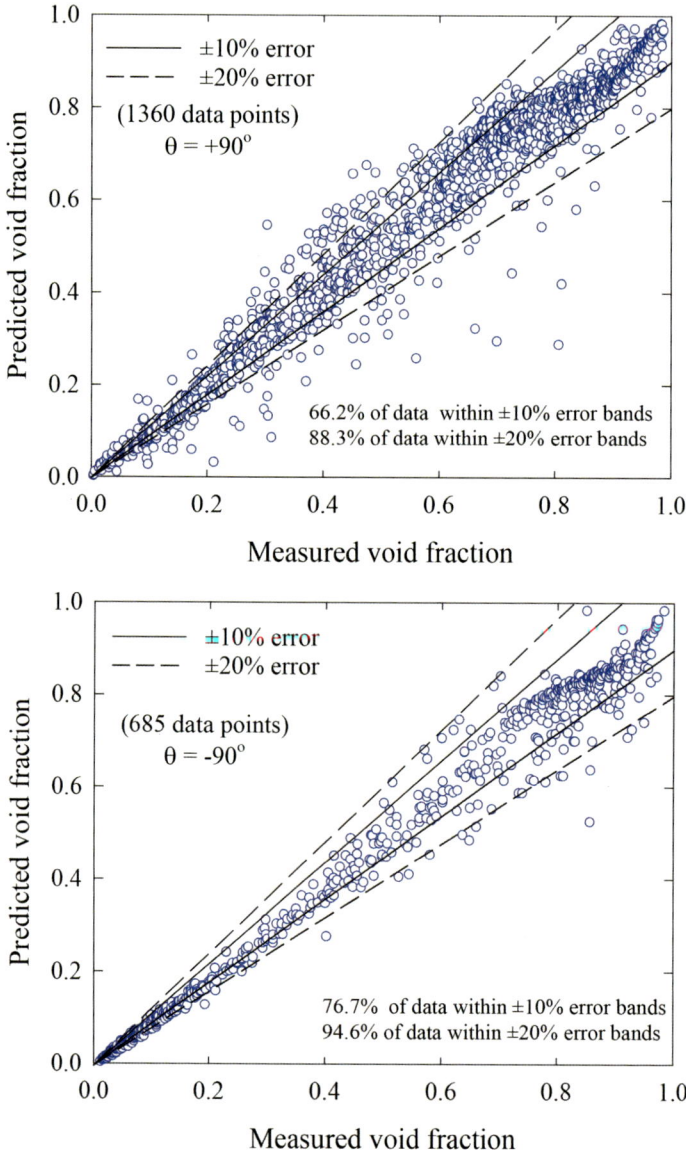

Fig. 4.13 Performance of the Bhagwat and Ghajar (2014) void fraction correlation against air–water data for vertical upward and downward pipe inclinations. (Data from the Two-Phase Flow Lab, OSU, Stillwater, OK)

depicted in Figs. 4.14 and 4.15. It is seen that the performance of their correlation deteriorates for downward pipe inclinations. The inaccuracies involved in prediction of void fraction in downward inclinations are essentially due to the flow stratification that causes the slip ratio to decrease ($\alpha_G > \lambda$). It is found that all drift flux and

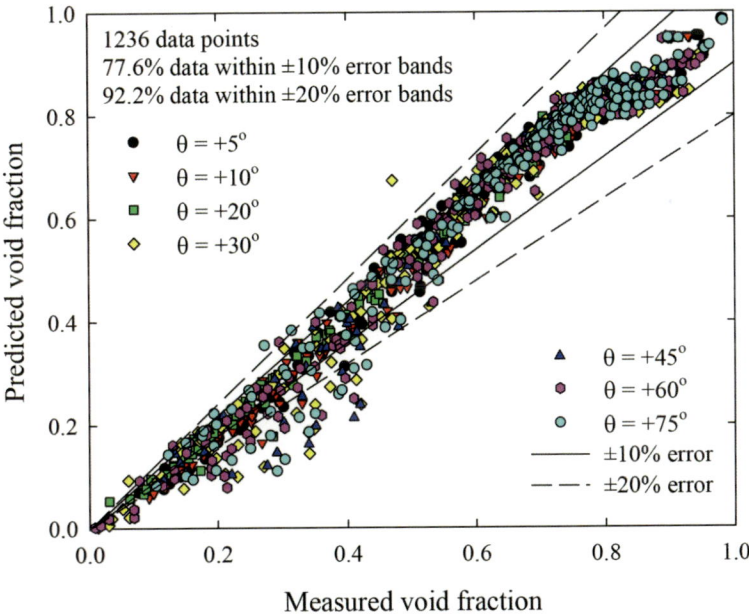

Fig. 4.14 Performance of the Bhagwat and Ghajar (2014) void fraction correlation against air–water data for upward inclined pipe inclinations. (Data from the Two-Phase Flow Lab, OSU, Stillwater, OK)

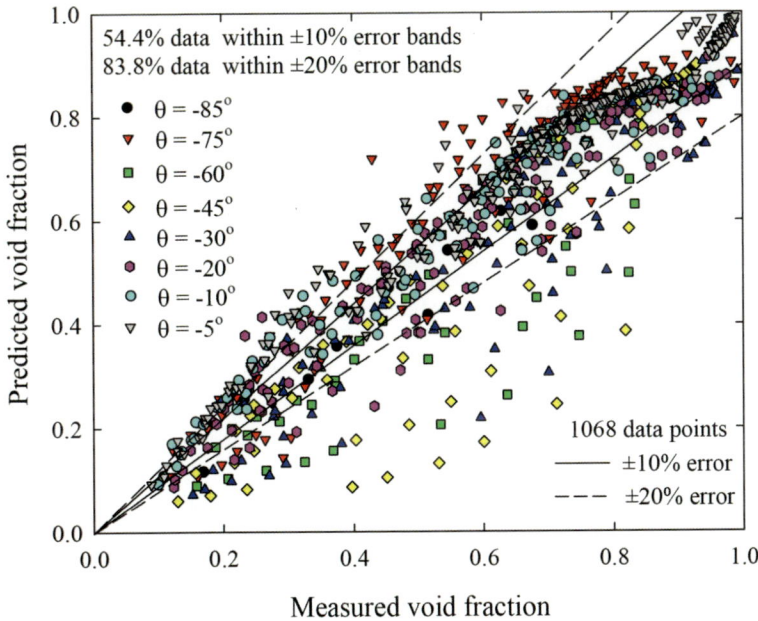

Fig. 4.15 Performance of the Bhagwat and Ghajar (2014) void fraction correlation against air–water data for downward inclined pipe inclinations. (Data from the Two-Phase Flow Lab, OSU, Stillwater, OK)

separated flow model-based correlations exhibit an underprediction tendency for all downward pipe inclinations when $\alpha_G > \lambda$.

The prediction of Bhagwat and Ghajar (2014) void fraction correlation against experimental void fraction data for different refrigerants is graphically presented in Fig. 4.16. For details about the sources of the data sets used, refer to Bhagwat and Ghajar (2014). With the exception of R22 and R134a, for all other refrigerants, their void fraction correlation predicts majority of the data within $\pm 20\%$ error bands. It should be noted that the data points for refrigerant R22 that deviate significantly from predicted values of Bhagwat and Ghajar (2014) correlation belong to the data of Sacks (1975) whose uncertainty analysis is not available. Other void fraction correlations available in the literature predicted these data points of R22 with more or less same error. In case of R134a, the overpredicted data belong to the experimental work of Bowers and Hrnjak (2010) who measured the void fraction in pipe diameters in a range of 7–15 mm through the pixel analysis of the images obtained by high-speed visualization. The reliability of these measurements needs further examination since they did not report uncertainty of the measured void fraction data and did not make any comparison regarding its agreement with other measurement techniques or with the predictions of available void fraction correlations. Bhagwat and Ghajar (2014) considered these data points of R22 and R134a as outliers however; these data points were included in the figures and performance analysis to avoid any bias in the experimental data used for the development of their void fraction correlation.

Figure 4.17 shows the excellent predictions of void fraction by Bhagwat and Ghajar (2014) void fraction correlation in mini and micro channels. Their correlation predicts the R410A experimental data of Shedd (2010) for pipe diameters in a range of 0.5–2.96 mm well within $\pm 20\%$ error bands. A close observation shows that for $0.75 < \alpha_G < 1$, their correlation predicts 79% of data within $\pm 5\%$ error bands.

Bhagwat and Ghajar (2014), in order to verify the ability of their void fraction correlation to handle high system pressures, compared its outcome with the measured void fraction data for high system pressures as shown in Fig. 4.18. Their correlation predicts the void fraction within $\pm 7.5\%$ of the measured values for different mass flux at high system pressure of 7.2 MPa and 18.1 MPa, respectively. For 7.2 MPa system pressure, their correlation predicts 76% of data points of Inoue et al. (1995) within $\pm 10\%$ error bands. In the case of 18.1 MPa system pressure, their correlation is found to predict 92% of data points of Liu et al. (2013) within $\pm 10\%$ error bands.

The performance of the Bhagwat and Ghajar (2014) void fraction correlation against air–oil void fraction data for different liquid dynamic viscosities is shown in Fig. 4.19. It is seen that for a wide range of liquid-phase dynamic viscosities, their correlation predicts majority of the data within $\pm 20\%$ error bands. For details about the sources of the data sets used, refer to Bhagwat and Ghajar (2014).

Since the expression for drift velocity (U_{GM}) used by Bhagwat and Ghajar (2014) as given in Eq. (4.16) incorporates a correction factor (c_3) to account for the reduced wall drag for large diameter pipes, it is also of interest to see how well their void fraction correlation predicted the void fraction data for large diameter pipes corresponding to $La < 0.025$. Based on the Laplace variable, the pipe diameters

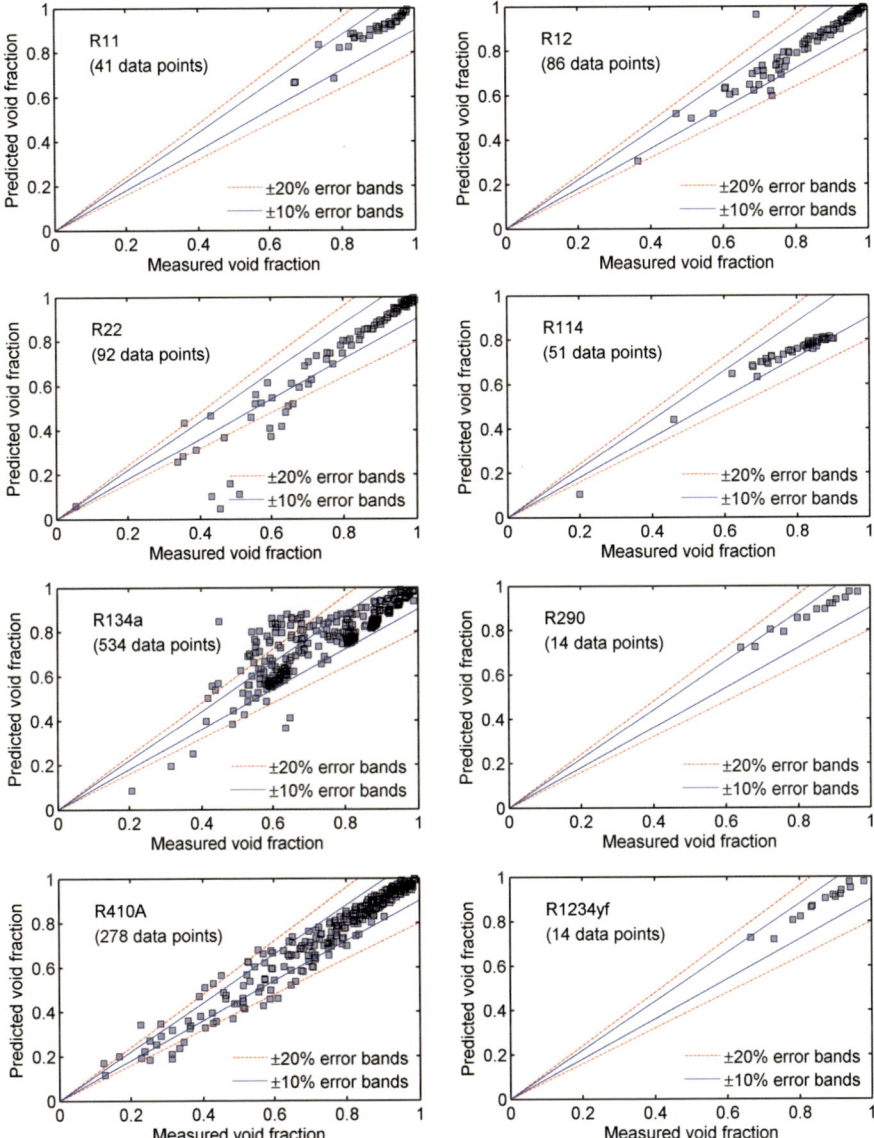

Fig. 4.16 Performance of the Bhagwat and Ghajar (2014) void fraction correlation against refrigerant void fraction data (0.5 mm $\leq D_h \leq$ 650 mm)

that classify as large diameter pipes are in a range of 0.067–0.305 m. The experimental data for large diameter pipes is obtained from Abdulkadir et al. (2010) for air–oil fluid combination and from Schlegel et al. (2010), Hills (1976), Inoue (2001), Shoukri et al. (2003), and Hashemi et al. (1986) for air–water fluid combination in vertical upward two-phase flow. With the exception of very few outliers for the data

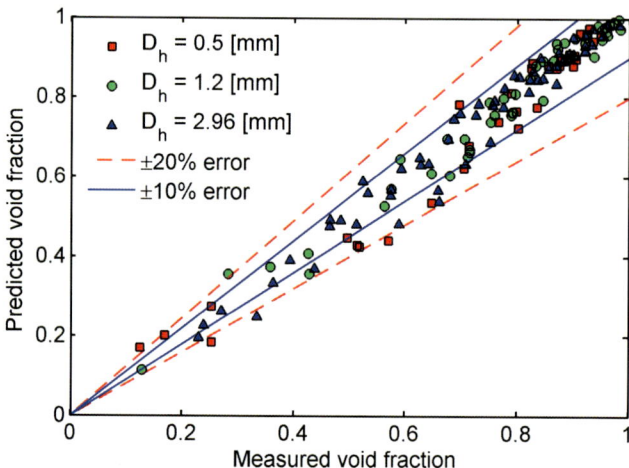

Fig. 4.17 Performance of the Bhagwat and Ghajar (2014) void fraction correlation against void fraction data of Shedd (2010) in mini and micro size channels using R410A

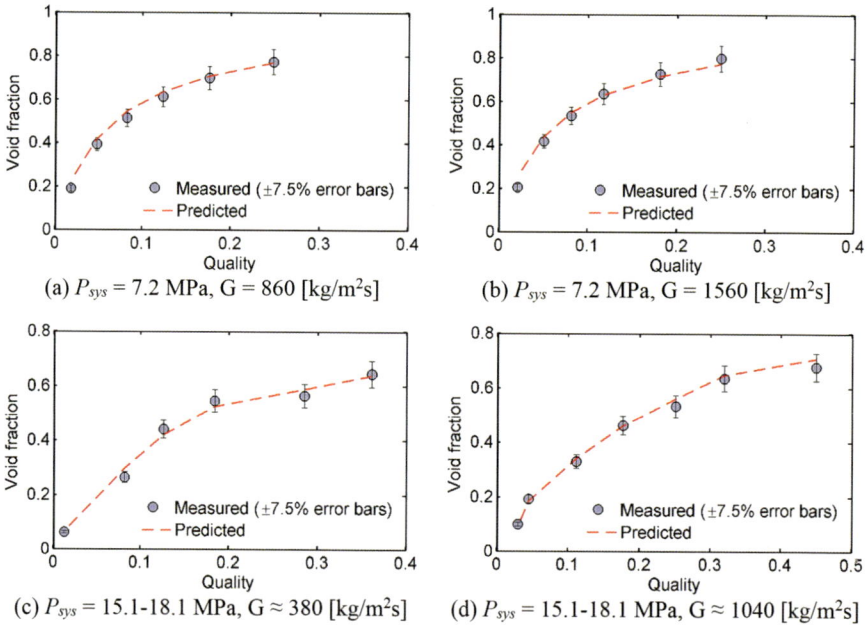

Fig. 4.18 Performance of the Bhagwat and Ghajar (2014) void fraction correlation against high-pressure steam–water void fraction data. (Experimental data for (**a**) and (**b**) from Inoue et al. (1995) and (**c**) and (**d**) from Liu et al. (2013))

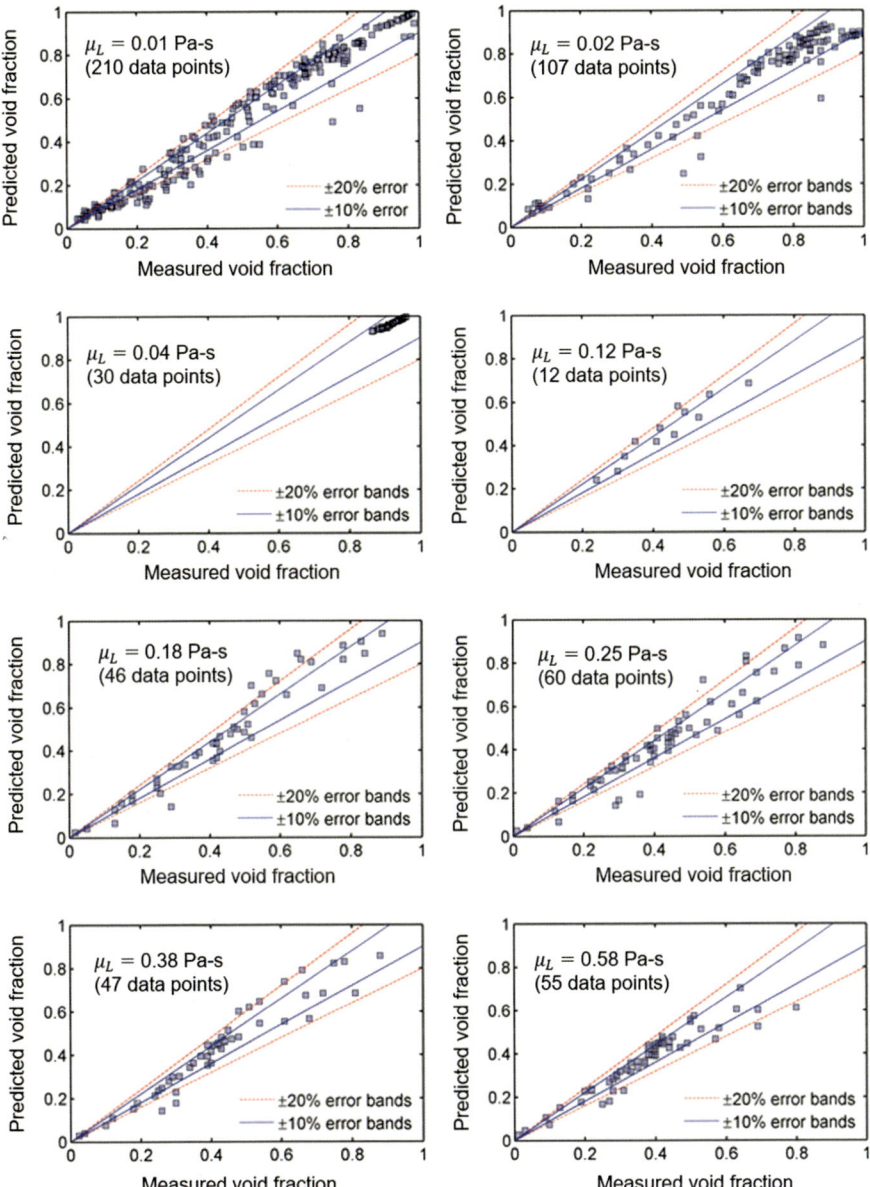

Fig. 4.19 Performance of the Bhagwat and Ghajar (2014) void fraction correlation against air–oil void fraction data (0.025 m $\leq D_h \leq$ 0.19 m)

of Hills (1976) and Shoukri et al. (2003), the void fraction correlation of Bhagwat and Ghajar (2014) predicts the void fraction data for $La < 0.025$ well within $\pm20\%$ error bands as shown in Fig. 4.20. The pipe diameter of 0.067 m for $La = 0.022$

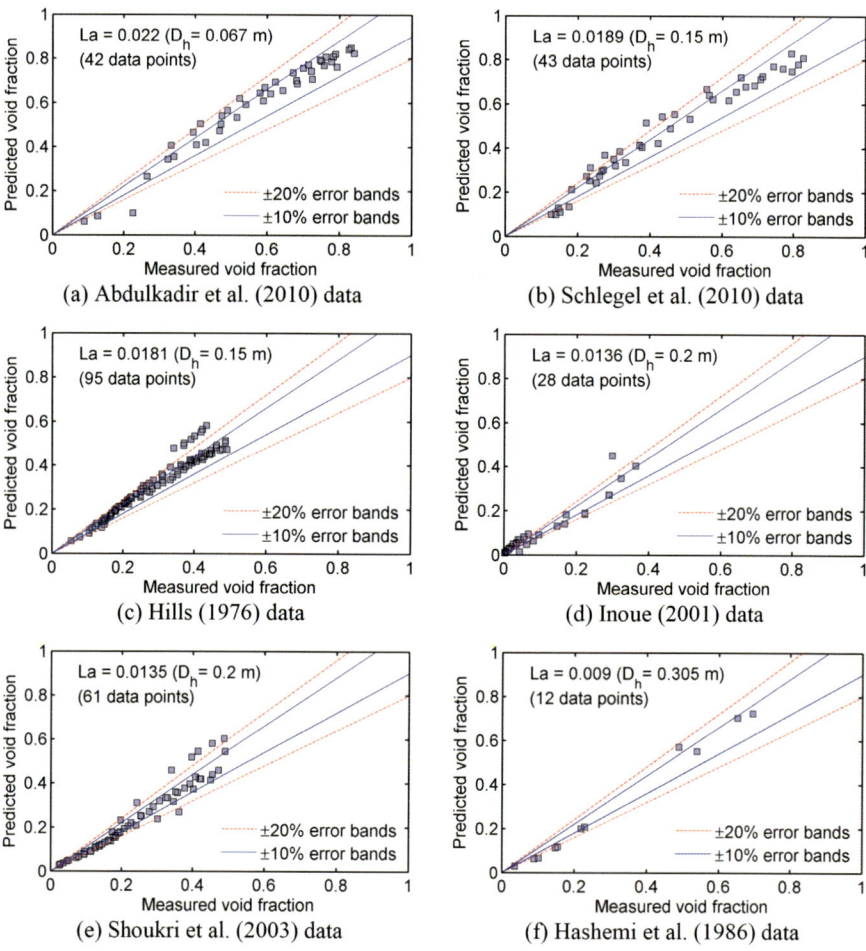

Fig. 4.20 Performance of the Bhagwat and Ghajar (2014) void fraction correlation against void fraction data in large diameter pipes ((**a**) air–oil data and (**b–f**) air–water data)

corresponds to air–oil data, while all other data sets belong to air–water fluid combination in vertical upward pipe orientation.

In addition to the large diameter pipes, the void fraction correlation of Bhagwat and Ghajar (2014) is also valid for noncircular pipe geometries. The performance of their correlation against the void fraction data in annular and rectangular geometries is illustrated in Fig. 4.21. Their correlation for the distribution parameter (C_o) given by Eq. (4.13) recommends use of Eq. (4.14) with constant $c_1 = 0.4$ for rectangular and annular ducts. The void fraction data for annular pipes belongs to air–water data (123 data points from Wongwises and Pipathttakul (2006) for $D_o = 12.5$ mm, $D_i = 8$ mm and 103 data points from Das et al. (2002) for $D_o = 50.8$ mm, $D_i = 25.4$ mm where D_o and D_i are the outer and inner pipe diameter of annulus,

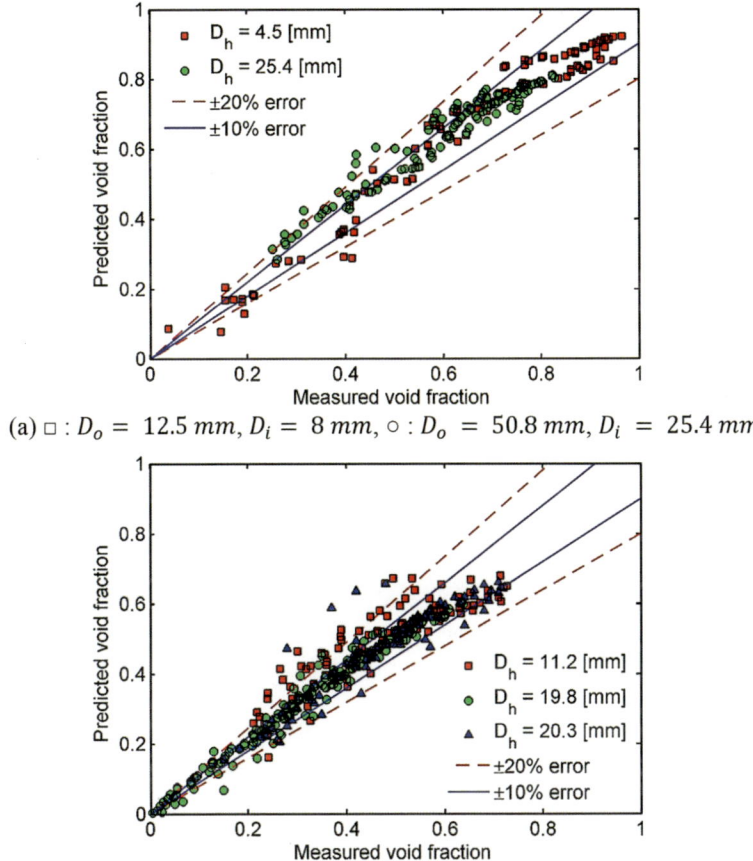

(a) □ : $D_o = 12.5\ mm$, $D_i = 8\ mm$, ○ : $D_o = 50.8\ mm$, $D_i = 25.4\ mm$

(b) □ : 6.35 mm × 50.8 mm, ○ : 11.1 mm × 93.6 mm and Δ : 12.7 mm × 50.8 mm

Fig. 4.21 Performance of the Bhagwat and Ghajar (2014) void fraction correlation against void fraction data in noncircular geometries ((**a**) annular channels and (**b**) rectangular channels)

respectively). The void fraction data in rectangular channels belong to steam–water data (157 data points from Marchaterre (1956) for 11.1 mm × 93.6 mm, 101 data points from Marchaterre et al. (1960) for 6.35 mm × 50.8 mm, and 98 data points from Marchaterre et al. (1960) for 12.7 mm × 50.8 mm rectangular channels).

The performance of the void fraction correlation of Bhagwat and Ghajar (2014) was also compared with some of the existing top performing void fraction correlations based on the drift flux and separated flow models. As detailed in Bhagwat and Ghajar (2014), their correlation consistently gave better performance over the entire range of parameters shown in Table 4.2 and the entire range of void fraction ($0 < \alpha_G < 1$) without any reference to flow regime map. It is evident from Table 4.2 that their correlation for void fraction covers pretty much the entire range of two-phase flow parameters that would exist in practical applications.

Note that the correlation of Bhagwat and Ghajar (2014) is not validated against data for micro channels with pipe diameters less than 0.5 mm. For gas–liquid two-phase flow in small size pipes less than 0.5 mm, correlation of Xiong and Chung (2006) could be used. Their correlation is based on the experimental data for pipe diameters in a range of 0.1–0.6 mm and consists of bubbly, slug, annular flow and entire range of gas volumetric flow fraction (λ). Their void fraction correlation, Eqs. (4.20) and (4.21), accounts for the nonlinear relationship between void fraction and gas volumetric flow fraction as a function of hydraulic pipe diameter (D_h) measured in mm:

$$\alpha_G = \frac{a_1 \lambda^{0.5}}{1 - (1 - a_1)\lambda^{0.5}} \left.\right\} \quad \begin{array}{l} 0.1 \le D_h \le 0.6 \text{ mm} \\ 4 \le Re_{SL} \le 1670, \;\; 4 \le Re_{SG} \le 1725 \end{array} \qquad (4.20)$$

$$a_1 = \frac{0.266}{1 + 13.8 \exp\left(-6.88 D_h\right)} \qquad (4.21)$$

It must be noted that drift flux model-based void fraction correlations are not suitable to model stratified type of flow. Compared to drift flux models, separated flow models typically work better for stratified flow. However, these models too cannot provide accurate prediction of the void fraction with desired accuracy. Stratified flow pattern in horizontal and downward pipe inclinations has a peculiar flow structure, and it is practically difficult to accurately model it with existing flow pattern-independent two-phase flow models. Apparently, the stratified flow pattern needs to be modeled using mechanistic flow model approach introduced by Taitel and Dukler (1976) and its derivatives. In addition to Taitel and Dukler (1976) model, two-phase literature also reports semi-mechanistic models called "apparent rough surface model" first introduced by Hamersma and Hart (1987) and Hart et al. (1989) and "double circle model" proposed by Chen et al. (1997), to calculate void fraction and two-phase pressure drop in stratified two-phase flow. It must be mentioned that the apparent rough surface and double circle models are developed only for stratified flow in horizontal pipe orientation and needs to be scrutinized against stratified flow data in downward pipe inclinations. For a detailed discussion on these three models for stratified flow, refer to Ghajar and Bhagwat (2017).

An Illustrative Example: Use of Woldesemayat and Ghajar (2007) and Bhagwat and Ghajar (2014) Void Fraction Correlations Consider the vertical upward two-phase flow of air and silicone oil in a 12 mm I.D. stainless steel pipe having a surface roughness of 0.002 mm. The mass flow rate of gas and liquid phase is 0.0015 kg/s and 0.9 kg/s, respectively. The fluid thermophysical properties may be taken as follows: $\rho_G = 1.2 \text{ kg/m}^3, \rho_L = 920 \text{ kg/m}^3, \mu_L = 0.005 \text{ Pa} \cdot \text{s}$, and $\sigma = 0.02$ N/m. The measured value of void fraction (α_G) for this flow is 0.5. Compare the measured value of α_G with the predicted values of α_G using (1) the explicit void fraction correlation of Woldesemayat and Ghajar (2007) and (2) the implicit void fraction correlation of Bhagwat and Ghajar (2014).

Solution

1. Calculation of α_G using Woldesemayat and Ghajar (2007) explicit void fraction correlation, given by Eqs. (4.11) and (4.12):

Referring to Table 1.1, the superficial gas (U_{SG}) and liquid (U_{SL}) velocities and the two-phase gas volumetric flow fraction (λ) are first calculated as follows:

$$U_{SG} = \frac{4\dot{m}_G}{\pi D^2 \rho_G} = \frac{4 \times 0.0015}{\pi \times 0.012^2 \times 1.2} = 11.05 \ \text{m/s}$$

$$U_{SL} = \frac{4\dot{m}_L}{\pi D^2 \rho_L} = \frac{4 \times 0.9}{\pi \times 0.012^2 \times 920} = 8.65 \ \text{m/s}$$

$$\lambda = \frac{U_{SG}}{U_{SG} + U_{SL}} = \frac{11.05}{11.05 + 8.65} = 0.561$$

Next, calculate the distribution parameter (C_o) using Eq. (4.11) and the drift velocity (U_{GM}) using Eq. (4.12). In using Eq. (4.12), assume $p_{atm} = p_{sys}$:

$$C_o = \lambda \left(1 + \left[\frac{U_{SL}}{U_{SG}} \right]^{(\rho_G/\rho_L)^{0.1}} \right) = (0.561)\left(1 + \left[\frac{8.65}{11.05} \right]^{(1.2/920)^{0.1}} \right) = 1.0556$$

$$U_{GM} = 2.9 \left[\frac{gD\sigma(1 + \cos\theta)(\rho_L - \rho_G)}{\rho_L^2} \right]^{0.25} (1.22 + 1.22 \sin\theta)^{p_{atm}/p_{sys}}$$

$$= 2.9 \left[\frac{(9.81)(0.012)(0.02)(1 + \cos 90)(920 - 1.2)}{920^2} \right]^{0.25} (1.22 + 1.22 \sin 90) = 0.283$$

Finally, the void fraction (α_G) is calculated from Eq. (4.1) where

$$\alpha_G = \frac{U_{SG}}{C_o U_M + U_{GM}} = \frac{11.05}{1.0556 \ (19.7) + (0.283)} = 0.52$$

where $U_M = U_{SL} + U_{SG} = 8.65 + 11.05 = 19.7$ m/s

2. Calculation of α_G using Bhagwat and Ghajar (2014) implicit void fraction correlation, given by Eqs. (4.13, 4.14, 4.15, 4.16, 4.17, 4.18, and 4.19):

To calculate the distribution parameter (C_o) using Eq. (4.13), the following calculations should first take place:

Referring to Table 1.1, the two-phase quality (x) is

$$x = \frac{\dot{m}_G}{\dot{m}_G + \dot{m}_L} = \frac{0.0015}{0.0015 + 0.9} = 0.00166$$

From Eq. (4.15), $Re_M = 4.35 \times 10^4$, and from the following equation for the Colebrook (1939) fanning friction factor correlation with $\varepsilon/D = 1.67 \times 10^{-4}$, the two-phase fanning friction factor is $f_M = 0.0055$:

$$\frac{1}{\sqrt{f_M}} = -4.0 \, \log \left(\frac{\varepsilon/D}{3.7} + \frac{1.256}{Re_M \sqrt{f_M}} \right)$$

Now, with $c_1 = 0.2$ for circular pipes and $\lambda = 0.561$ (see part 1 of this problem), the value of $C_{o,1}$ from Eq. (4.14) can be determined.

To calculate U_{GM} from Eq. (4.16), the following calculations should first take place:

From Eq. (4.17), since $\mu_L/0.001 \leq 10 \Rightarrow c_2 = 1$

From Eqs. (4.19) and (4.18), since $La = 0.124 \geq 0.025 \Rightarrow c_3 = 1$

Now, from Eq. (4.13) for the distribution parameter (C_o) and Eq. (4.16) for the drift velocity (U_{GM}), the iterative calculations yield, $C_o = 1.2$ and $U_{GM} = 0.088$ m/s. Finally, the void fraction (α_G) is calculated from Eq. (4.1) where

$$\alpha_G = \frac{U_{SG}}{C_o U_M + U_{GM}} = \frac{11.05}{1.2 \, (19.7) + (0.088)} = \mathbf{0.47}$$

where $U_M = U_{SL} + U_{SG} = 8.65 + 11.05 = 19.7$ m/s

Discussion The explicit void fraction correlation of Woldesemayat and Ghajar (2007), Eqs. (4.11) and (4.12), overpredicted the measured value of α_G by 4% (0.52 vs 0.5). The implicit void fraction correlation of Bhagwat and Ghajar (2014), Eqs. (4.13, 4.14, 4.15, 4.16, 4.17, 4.18, and 4.19), underpredicted the measured value of α_G by 6% (0.47 vs 0.5). Both correlations did an excellent job of predicting the experimental value of void fraction for this flow.

Chapter 5
Pressure Drop

Background

Similar to single-phase flow, total pressure drop in gas–liquid two-phase flow is comprised of hydrostatic, frictional, and accelerational components as was shown in Eq. (2.1). However, unlike the single-phase flow, these components of two-phase pressure drop in gas–liquid flows may depend upon several parameters such as void fraction, pipe orientation, flow patterns, and liquid entertainment fraction. The hydrostatic or gravitational component of two-phase pressure drop due to the pipe inclination depends on the two-phase mixture density which in turn is a function of flow patterns and void fraction. Accelerational pressure drop is due to the change in momentum of the two-phase mixture and can be neglected for non-boiling (adiabatic) two-phase flows within a short pipe length. However, for two-phase flow undergoing phase change process or even for adiabatic two-phase flow in long and inclined pipes, depending upon the flow pattern, the accelerational pressure drop can contribute up to 20% of the total two-phase pressure drop. Finally, the frictional component of two-phase pressure drop is a result of the interfacial friction between the two phases and wall friction with both or either phase and depends upon the flow pattern structure, pipe diameter, and fluid properties.

The two-phase pressure drop is influenced by several parameters such as pipe orientation, flow rates of individual phases, pipe diameter, fluid properties, and surface roughness. These influences on the two-phase pressure drop will be discussed next.

© The Author(s), under exclusive license to Springer Nature Switzerland AG 2020
A. J. Ghajar, *Two-Phase Gas-Liquid Flow in Pipes with Different Orientations*,
SpringerBriefs in Applied Sciences and Technology,
https://doi.org/10.1007/978-3-030-41626-3_5

Effect of Pipe Orientation on Pressure Drop

The effect of pipe orientation on total two-phase pressure drop is primarily due to the effect of pipe orientation on the hydrostatic component of the two-phase pressure drop. Two-phase literature reports very few studies focused on measurement of two-phase pressure drop over the entire range of pipe orientations. Some of these studies include work of Beggs (1972), Mukherjee (1979), Spedding et al. (1982), and Lips and Meyer (2012). The total two-phase pressure drop measurements for different pipe orientations ranging from +90° to −90° carried out at the Two-Phase Flow Lab, OSU, and that reported by Lips and Meyer (2012) are presented in Figs. 5.1 and 5.2, respectively.

A comparison between Figs. 5.1 and 5.2 shows that the variation of the total two-phase pressure drop as a function of pipe orientation is similar for both non-boiling (Fig. 5.1) and condensing (Fig. 5.2) two-phase flows. It is evident that at low gas and liquid flow rates, the total two-phase pressure drop is relatively insensitive to the change in downward pipe inclination. This is due to the fact that at these flow rates, stratified flow exists, and its physical structure and associated void fraction are relatively insensitive to the change in pipe orientation. In upward pipe inclinations, the total two-phase pressure drop increases rapidly with increase in the pipe orientation, and it is fascinating to see that the total two-phase pressure drop at lower gas flow rates is greater than that at higher gas flow rates. This is because of the reversal of liquid film that induces decreasing trend of pressure gradient with increase in gas flow rates. For a detailed discussion on the pressure gradient minimum and flow reversal in upward inclined flow, refer to Ghajar and Bhagwat (2017). At higher liquid flow rates, negative values of total two-phase pressure drop in downward pipe inclinations represent partial pressure recovery due to gain in hydrostatic component of two-phase pressure drop.

Effect of Phase Flow Rates on Pressure Drop

The effect of change in phase flow rates on the two-phase pressure drop is essentially due to the change in physical structure of the flow patterns and is of different nature for the hydrostatic and frictional components of two-phase pressure drop. Since the hydrostatic component of two-phase pressure drop depends only on the void fraction (see Eq. 2.2), for a fixed liquid flow rate, increase in gas flow rate decreases the contribution of hydrostatic pressure drop (due to decrease in mixture density) where it increases the frictional pressure drop component. The relation between change in phase flow rates and the frictional pressure drop could be construed in a better way by presenting their relationship in different coordinate systems. As shown in Fig. 5.3, for low liquid and low gas flow rates, the two-phase frictional pressure drop remains virtually constant and increases drastically only for moderate to high gas flow rates (intermittent and annular flow regimes). For intermittent type of flow,

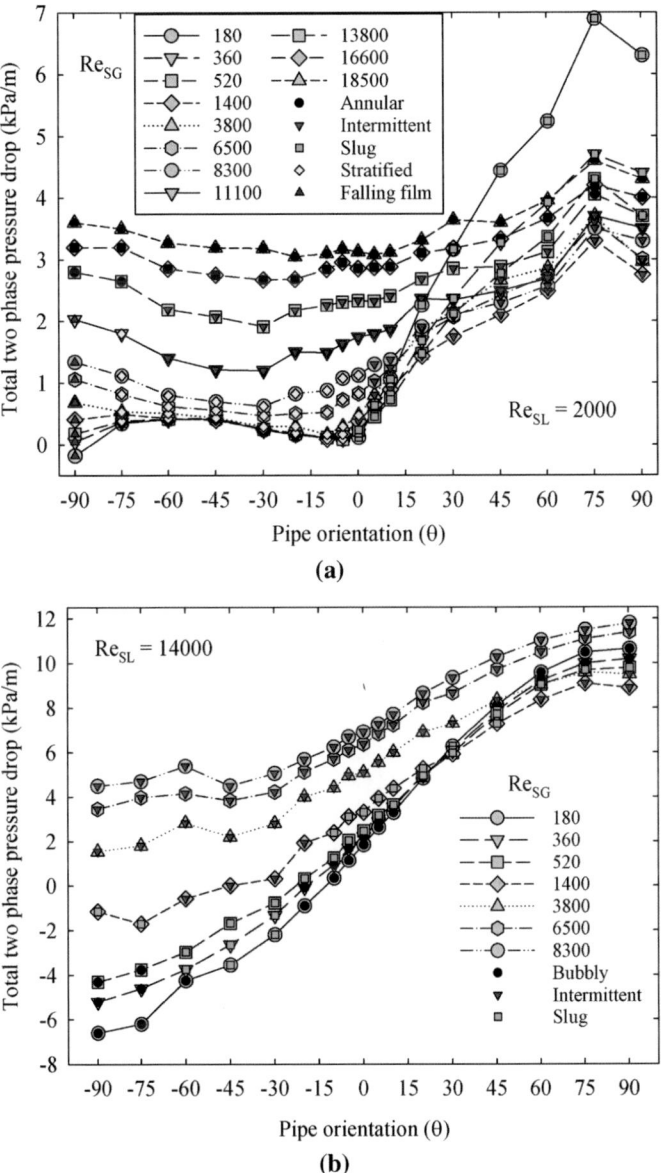

Fig. 5.1 Effect of pipe orientation on total two-phase pressure drop. (**a**) $Re_{SL} = 2000$ and $Re_{SG} = 180 - 18,500$. (**b**) $Re_{SL} = 14,000$ and $Re_{SG} = 180 - 8300$. (Data measured at the Two-Phase Flow Lab, OSU, Stillwater, OK)

Fig. 5.2 Effect of pipe orientation on two-phase pressure drop. (Data of Lips and Meyer 2012)

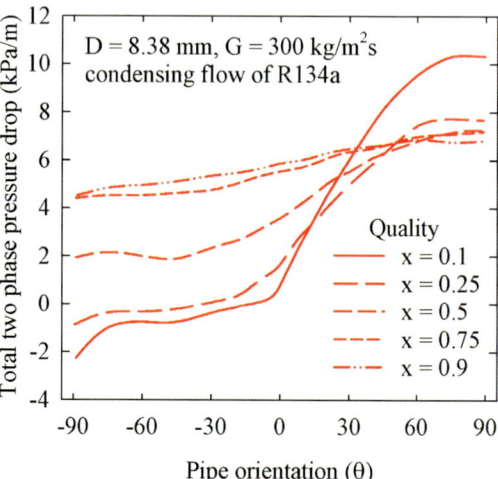

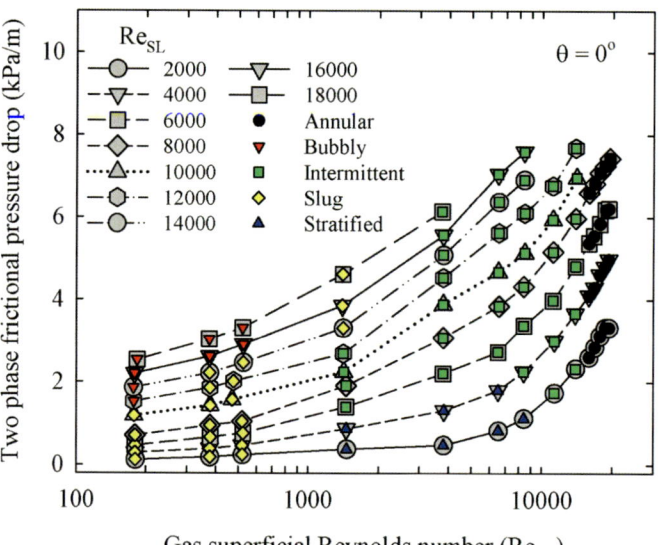

Fig. 5.3 Two-phase frictional pressure drop for varying gas and liquid flow rates. (Data measured at the Two-Phase Flow Lab, OSU, Stillwater, OK)

sharp increase in the frictional pressure drop is essentially due to the turbulent and chaotic nature of the two-phase mixture. For annular flow, the pressure gradient increases sharply with increase in the gas and liquid flow rates. At a fixed gas flow rate, increase in the liquid flow rate increases the liquid film thickness and thus reduces the cross-sectional area available for the gas phase. This increases the actual gas velocity and exerts higher shear on the gas–liquid interface to increase the two-phase frictional pressure drop. Additionally, the gas–liquid interface offers a

Fig. 5.4 Two-phase
pressure drop as a function
of phase flow rates. (Data of
Quiben and Thome 2007)

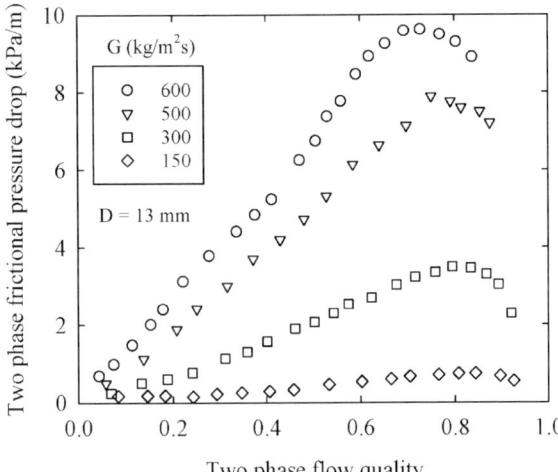

rough surface to the gas flow, and the interface gets roughened progressively with
the increase in the liquid film thickness or alternatively the liquid flow rate that
augments the frictional pressure drop.

Variation of two-phase pressure drop with change in two-phase flow rates could
also be presented in another perspective using two-phase flow quality data of Quiben
and Thome (2007). This type of presentation shown in Fig. 5.4 is usually adopted for
high-pressure systems (often encountered in refrigeration and nuclear applications)
which tend to occupy the entire range of two-phase flow quality ($0 < x < 1$). It is
evident that the two-phase pressure drop increases systematically with increase in the
mass flux (G) and quality (x). An inflection point that flips the trend of increase in
two-phase pressure drop with increase in quality is quite noticeable. This change in
the trend of two-phase pressure drop could be explained based on the change in
structure of the two-phase flow pattern at the inflection point. By the time the
two-phase flow system reaches the point of maximum pressure drop (inflection
point), the flow pattern attains annular flow structure; however, beyond that point
due to severe entrainment process, the liquid phase in contact with the pipe wall is
gradually moved to the central gas core reducing the liquid film thickness and
consequently reducing the effective viscosity in the near-wall region. This ultimately
results into a reduced frictional pressure drop. This type of trend is not evident in
Fig. 5.3 since the data presented there does not contain two-phase flow with
significant entrainment. It is also evident from Fig. 5.4 that the quality at the point
of maximum pressure drop shifts toward lower qualities with increase in the
two-phase mixture mass flux. Additionally, Ducoulombier et al. (2011) reported
that quality associated with maximum frictional pressure drop also depends upon the
fluid properties. The combined effect of mass flux and fluid thermophysical properties on the quality at maximum pressure drop is reported in Fig. 5.5. It is clear that the
two-phase flow quality corresponding to the point of maximum pressure drop may
vary approximately between 0.65 and 0.95.

Fig. 5.5 Influence of mass flux and fluid properties on quality corresponding to maximum pressure drop. (Adapted from Ducoulombier et al. 2011)

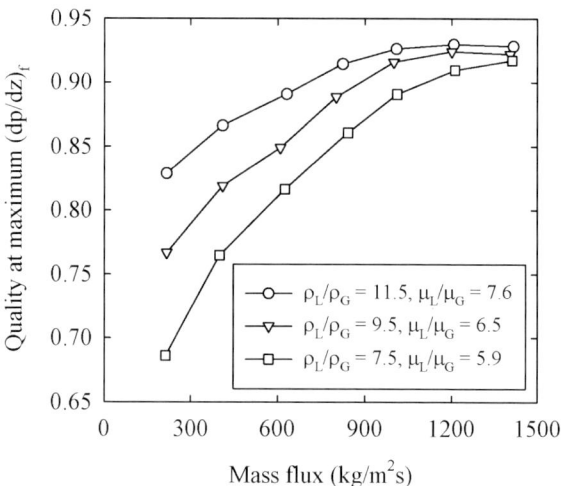

Effect of Pipe Diameter on Pressure Drop

Similar to the single-phase pressure drop, the effect of decrease in pipe diameter is to increase the two-phase frictional pressure drop. However, this effect of pipe diameter is usually in conjunction with two-phase flow patterns. The two-phase frictional pressure drop in annular flow regime depends on the pipe diameter to a great extent where it is relatively less sensitive to the pipe diameter in bubbly flow regime. It is evident from Fig. 5.6a that for a fixed mass flux of R134a, the two-phase frictional pressure drop for different pipe diameters deviates significantly for higher values of two-phase flow quality (annular flow pattern) whereas for low values of two-phase flow quality (bubbly flow), the effect of pipe diameter on two-phase frictional pressure drop gradually diminishes. Similar conclusions could be drawn for air–water two-phase flow system from the work of Kaji and Azzopardi (2010) who studied the effect of pipe diameters in a range of 10–50 mm on two-phase pressure drop in annular flow regime. As shown in Fig. 5.6b, for a fixed liquid flow rate, the effect of change in pipe diameter is most significant at higher gas flow rates (i.e., higher two-phase flow qualities). The two-phase frictional pressure drop is due to the friction at the pipe wall as well as the gas–liquid interface, and the gas–liquid interfacial area increases with increase in the pipe diameter. Thus, it is evident that any two-phase frictional pressure drop correlation must definitely account for the pipe diameter effect in annular flow regime.

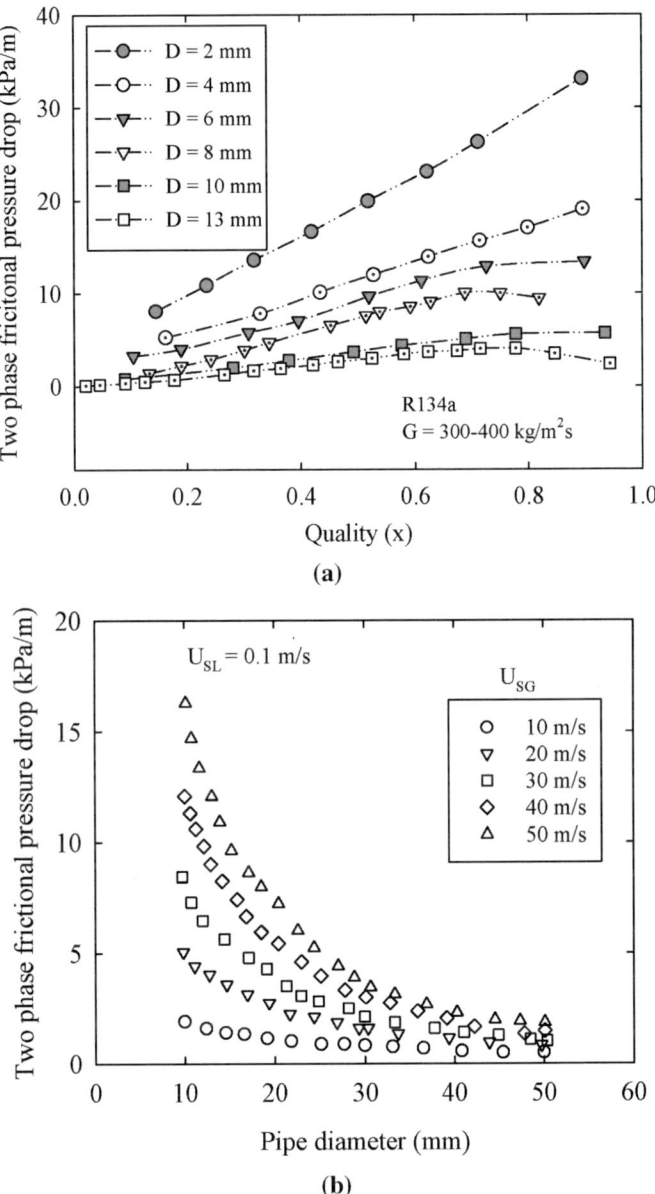

Fig. 5.6 Effect of pipe diameter on two-phase frictional pressure drop. (**a**) Two-phase flow of R-134a refrigerants. (**b**) Two-phase flow of air-water. (Adapted from Kaji and Azzopardi 2010)

Effect of Fluid Properties on Pressure Drop

The important fluid properties that show a noticeable effect on two-phase frictional pressure drop are the gas-phase density and liquid-phase dynamic viscosity. The increase in gas-phase density decreases the slippage at the gas–liquid interface and hence decreases the frictional component of two-phase pressure drop, whereas the increase in liquid dynamic viscosity increases the shear in liquid phase in contact with the pipe wall and also the shear at gas–liquid interface resulting in increase in frictional pressure drop. Intuitively, it can be said that the effect of fluid properties on two-phase frictional pressure drop is more prominent for shear-driven flows compared to buoyancy-driven two-phase flow. For more details on the effect of fluid properties on frictional pressure drop, refer to Oshinowo (1971), Fukano and Furukawa (1998), Abduvayat et al. (2003), Hlaing et al. (2007), and Gokcal (2008).

Effect of Surface Roughness on Pressure Drop

The effect of pipe surface roughness on two-phase pressure drop is crucial in applications involving two-phase flow through steel or micro-finned tubes. The internally ribbed or micro-finned tubes are used in air conditioning and refrigeration applications to improve the tube side heat transfer, however, at the expense of enhanced pressure drop. Most of the two-phase flow research reported in the literature is carried out in a transparent (smooth) pipe, while the effect of pipe surface roughness on the frictional two-phase pressure drop is a relatively less investigated issue. Work of Wongs-ngam et al. (2004) and Shannak (2008) reported a considerable effect (20–60%) of wall roughness on pressure drop especially for the two-phase flow at high mass flux and quality. Figure 5.7 shows that the effect of pipe wall roughness on two-phase pressure drop is considerable at high gas velocities and increases with increase in liquid velocity. Obviously, similar to single-phase flow, the effect of wall roughness on two-phase pressure drop increases with decrease in

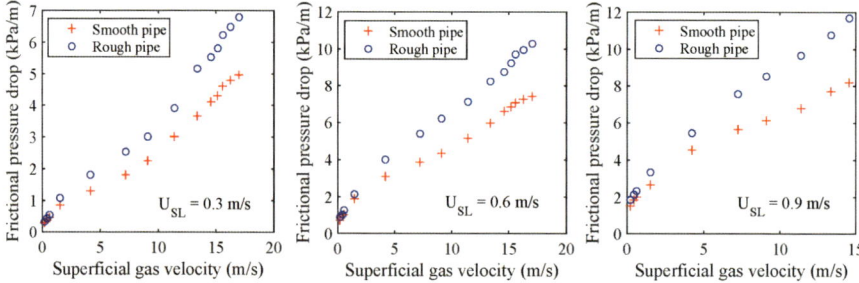

Fig. 5.7 Effect of pipe surface roughness on two-phase frictional pressure drop. (Data measured at the Two-Phase Flow Lab, OSU, Stillwater, OK) ($D = 12.5\,\text{mm}$, $\varepsilon = 20\mu m$ for rough pipe)

the pipe diameter. Experiments performed at the Two-Phase Flow Lab, OSU, confirm these observations. Moreover, their work also shows that the effect of pipe roughness on two-phase pressure drop is independent of the pipe orientation. The effect of pipe surface roughness on two-phase frictional pressure drop is maximum for annular flow (at high gas flow rates) since, in addition to the effect of pipe wall roughness acting on the liquid phase, the structure of this flow pattern also offers a continuous and rough gas–liquid interface responsible for enhanced pressure drop.

Pressure Gradient Minimum and Flow Reversal in Upward Inclined Flow

In concurrent gas–liquid upward inclined two-phase flow, at low liquid and moderate gas flow rates, although the net two-phase flow is observed in the upward direction, reversal of liquid film in contact with the pipe wall is observed under the influence of gravity forces. This phenomenon is a consequence of the interaction between the interfacial shear stress (exerted by gas phase on liquid film) and the gravitational forces acting on the liquid film. The liquid film in contact with the pipe wall may undergo a partial reversal (fluctuating wall shear stress) or complete reversal (negative wall shear stress) depending upon the gas and liquid flow rates and exhibits a decreasing trend of two-phase pressure gradient.

As long as there is a balance between interfacial shear stress and hydrostatic pressure drop, liquid film in contact with the pipe wall will either remain stationary or oscillate giving a net zero wall shear stress. Thus, when the interfacial shear stress is large enough to supersede the gravity forces (in form of hydrostatic pressure drop), the net flow of two-phase mixture will be in upward direction without any reversal of liquid film.

Experimentally, this condition could be determined by analyzing the pressure drop signal as a function of gas and liquid flow rates. At a fixed liquid flow rate and with an increase in gas flow rate, a point is reached where the gas phase does not have enough potential to carry liquid phase along with it in the downstream direction. At this point, the liquid phase in contact with the pipe wall appears to oscillate or move in the downward direction. The point corresponding to the inception of decreasing pressure gradient trend (in spite of increase in gas flow rate) is regarded as the *flow reversal point*, while the condition at which the pressure gradient commence to increase again with increase in gas flow rate is identified as *pressure gradient minimum point*. The two-phase literature provides several instances of decreasing pressure gradient trends in context to churn-annular transition in vertical upward two-phase flow. As a matter of fact, the decreasing trend of pressure gradient minimum also exists at lower gas flow rates during slug-churn transition in upward inclined pipes. Nevertheless, the flow physics governing these two separate regions of pressure gradient minimum is significantly different. A

Fig. 5.8 Non-dimensional
pressure drop for varying
non-dimensional gas and
liquid flow rates. (Data
measured at the Two-Phase
Flow Lab, OSU, Stillwater,
OK)

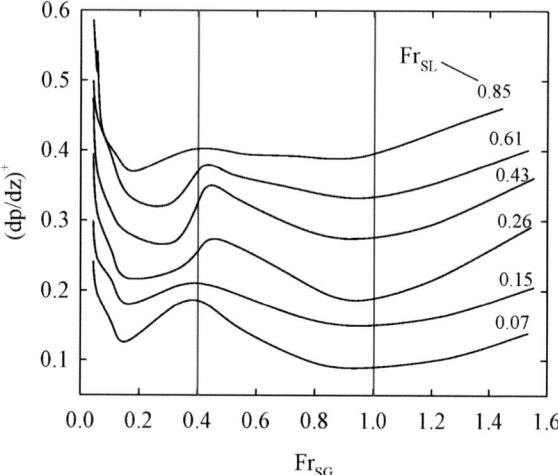

typical example of decreasing trends of pressure gradient minimum in vertical
upward pipe inclination is illustrated in Fig. 5.8. The measured two-phase pressure
drop and corresponding gas and liquid flow rates are made non-dimensional as
shown in Eqs. (5.1, 5.2, and 5.3), respectively:

$$\left(\frac{dp}{dz}\right)_{t}^{+} = \frac{(dp/dz)_{t}}{(\rho_{L} - \rho_{G})g} \tag{5.1}$$

$$Fr_{SG} = \frac{U_{SG}}{\sqrt{gD}}\sqrt{\frac{\rho_{G}}{(\rho_{L} - \rho_{G})}} \tag{5.2}$$

$$Fr_{SL} = \frac{U_{SL}}{\sqrt{gD}}\sqrt{\frac{\rho_{L}}{(\rho_{L} - \rho_{G})}} \tag{5.3}$$

It is seen from Fig. 5.8 that for a fixed liquid flow rate, when the gas flow rate is
increased, the two-phase pressure gradient exhibits first minimum at $Fr_{SG} \approx 0.2-0.3$
and a maximum at $Fr_{SG} \approx 0.4$. The first instance of decreasing pressure drop is due
to falling film surrounding the gas slug. As the gas slug rises in downstream
direction, it sheds liquid phase surrounding it to maintain the continuity. The sudden
increase in pressure drop between $Fr_{SG} \approx 0.2-0.4$ is due to the high level of
turbulence caused by disintegration of gas slug during slug to churn/intermittent
flow transition. Further this point of maximum pressure gradient at $Fr_{SG} \approx 0.4$,
churn/intermittent flow is known to commence, and again a decreasing trend of
pressure gradient followed by a pressure gradient minimum is observed at
$Fr_{SG} \approx 0.9-1$. Beyond this point, annular flow is known to exist. Note that this

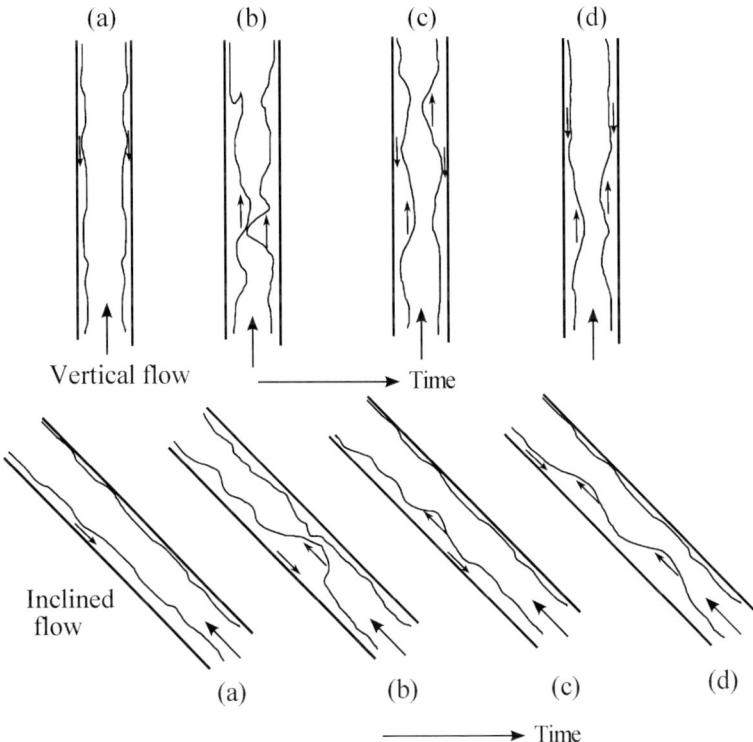

Fig. 5.9 Schematics of mechanism governing pressure gradient minimum phenomenon

value of the non-dimensional gas flow rate is essentially the criteria given by Eq. (3.25).

The second trend of decreasing pressure drop could be explained using Fig. 5.9. During churn-annular flow transition, large interfacial disturbance waves are generated that travel in downstream direction (Fig. 5.9-sketch b). During the swiping action of disturbance wave, liquid film travels in downstream direction under the influence of interfacial drag. However, once the disturbance wave passes by a certain pipe cross section, there is no driving potential for the liquid film, and it tends to fall back under the influence of gravity (Fig. 5.9 sketch c and d). The direction of travel of liquid film for different cases is depicted by small arrows inside the film. During and just after the encounter of liquid film with the disturbance wave, the portion of liquid film near interface may travel in upward direction, while that away from interface (or near pipe wall) may travel in downward direction leading to severe fluctuations in shear stress and velocity distributions in liquid film. The falling liquid film is again lifted up by upcoming disturbance wave, and the process continues until the frequency of disturbance waves is high enough to prevent reversal of liquid film. More details about this phenomenon and its dependency on disturbance waves could be found in Hewitt et al. (1965, 1985), Hewitt and Lacey (1965), and Owen (1986).

When it is desirable to operate a two-phase flow system in the region *not affected* by reversal of liquid film, it is crucial to identify the gas and liquid flow rates that belong to the region *affected* by flow reversal. Compared to the first trend of decreasing pressure gradient, its second instance is more important since this region is marked by severe turbulence and instability of the liquid film. The two-phase flow literature reports that the point of second pressure gradient minimum occurs at $Fr_{SG} \approx 1$. However, this criterion does not provide any idea on the range of liquid flow rates for which this trend would exist. Experiments carried out at the Two-Phase Flow Lab, OSU, reveal that the liquid flow rates at which this trend of second pressure gradient minimum would exist depend upon the pipe orientation such that the relationship between non-dimensional liquid flow rates and pipe orientation corresponding to second pressure gradient minimum is given by Eq. (5.4). Note that this relationship need not be used for pipe orientations less than $+10°$ from horizontal since the effect of pipe orientation on pressure gradient is negligibly small. For a fixed pipe orientation, Eq. (5.4) gives a threshold value of non-dimensional liquid flow rate (Fr_{SL}) above which decreasing pressure gradient trends or alternatively reversal of liquid film will no longer exist. Independent of pipe orientation and liquid flow rates, the non-dimensional gas flow rates (Fr_{SG}) corresponding to this point are in the range of 0.9–1:

$$Fr_{SL} = 0.6575(\sin\theta)^{1.1175} \left.\right\} \quad +10° \leq \theta \leq +90° \qquad (5.4)$$

Two-Phase Pressure Drop Modeling

As was shown in Eq. (2.1), the total pressure drop in gas–liquid two-phase flow consists of three main components: hydrostatic pressure drop (due to pipe elevation), pressure drop due to friction at the pipe wall and gas–liquid interface, and pressure drop due to acceleration (due to change in specific volume of the two-phase mixture). The contribution of each of these components to the total pressure drop depends upon the flow patterns, fluid properties, pipe orientation, and the pipe diameter. For example, in the case of bubbly flow in vertical upward orientation, the hydrostatic pressure drop component can be dominant in comparison to the frictional pressure drop depending upon the pipe diameter, while, for annular flow regime, the frictional pressure drop has a prominent share in the total pressure drop and the hydrostatic pressure drop component is negligible. Moreover, the contribution of the frictional pressure drop to the total pressure drop is observed to increase with decreasing pipe diameter. The total two-phase pressure drop per unit pipe length is the summation of its three components as expressed by Eq. (2.1). The right hand side of Eq. (2.1) is typically modeled using homogeneous and separated flow models. A brief description of these models will be presented next.

Homogeneous Flow Model The correlations based on homogenous flow model assume that the two phases are well mixed with each other and move with identical velocities (there is no slippage at the gas–liquid interface). The homogenous flow model thus considers the two-phase mixture as a pseudo single-phase mixture having two-phase physical properties. Under the consideration of homogeneous flow model, the hydrostatic two-phase pressure drop is calculated using homogeneous two-phase mixture density (ρ_M) as shown in Eq. (5.5). The two-phase mixture density can be expressed either in terms of two-phase flow quality (x) or the gas volumetric flow fraction (λ) as shown in Eq. (5.6). See Table 1.1 for interchangeability between λ and x:

$$\left(\frac{dp}{dz}\right)_h = \rho_M g \sin\theta \tag{5.5}$$

$$\rho_M = \left(\frac{x}{\rho_G} + \frac{1-x}{\rho_L}\right)^{-1} = \lambda\rho_G + (1-\lambda)\rho_L \tag{5.6}$$

The pressure drop due to acceleration of the two-phase mixture is calculated from Eq. (5.7). For the case of two-component two-phase flow (non-boiling, non-condensing), the two-phase flow quality remains constant for a relatively short length of pipe, and the magnitude of pressure drop due to change in specific volume of two-phase mixture is negligible and hence is ignored:

$$\left(\frac{dp}{dz}\right)_a = G^2 \frac{dv_M}{dz} = G^2 \left(v_{LG}\frac{dx}{dz} + x\frac{dv_G}{dp}\frac{dp}{dz}\right) \tag{5.7}$$

Finally, the frictional component of two-phase pressure drop is expressed in terms of the two-phase friction factor as shown in Eq. (5.8). The two-phase friction factor (f_M) is calculated based on two-phase Reynolds number (Re_M) which in turn is the function of two-phase mixture dynamic viscosity (μ_M) as shown in Eq. (5.9). Two-phase flow literature provides several models to calculate two-phase dynamic viscosity as a function of two-phase flow quality. For more details on the different two-phase dynamic viscosity models, refer to Ghajar and Bhagwat (2017):

$$\left(\frac{dp}{dz}\right)_f = \frac{2f_M G^2}{D\rho_M} \tag{5.8}$$

$$Re_M = \frac{GD}{\mu_M} \tag{5.9}$$

The two-phase friction factor (f_M) is calculated using single-phase fanning friction factor correlations such as Blasius (1913) and Churchill (1977).

Thus, the total two-phase pressure drop is the summation of the hydrostatic, accelerational, and frictional two-phase pressure drops as given by Eqs. (5.5), (5.7), and (5.8), respectively. For a more detailed discussion on the homogenous flow model, refer to Ghajar and Bhagwat (2017).

Separated Flow Model The correlations based on separated flow model assume the two phases to flow separately and share a definite and continuous interface between them, and unlike homogeneous flow model, it considers the slippage at the gas–liquid interface. Thus, using two- phase separated flow model, the two-phase mixture density (ρ_M) and hence the two-phase hydrostatic pressure drop calculation are based on the void fraction (α_G) as given by Eq. (5.10) where the two-phase mixture density is given by Eq. (5.11):

$$\left(\frac{dp}{dz}\right)_h = \rho_M g \sin\theta = (\alpha_G \rho_G + (1 - \alpha_G)\rho_L)g \sin\theta \tag{5.10}$$

$$\rho_M = \alpha_G \rho_G + (1 - \alpha_G)\rho_L \tag{5.11}$$

The accelerational component of two-phase pressure drop is expressed by Eq. (5.12). This expression is valid for all but annular flow pattern with considerable liquid entrainment. In particular, for annular flow, when the entrainment of liquid droplets into the central gas core is significant, the error associated with the assumption of negligible accelerational pressure drop can be severe. In the case of annular flow with liquid entrainment, the process of entrainment also contributes to the two-phase pressure drop and hence cannot be ignored. For such a specific case (annular flow with liquid entrainment), Eq. (5.12) can be expressed in the form of Eq. (5.13). See Chap. 6 for a detailed discussion on the entrainment mechanisms and the different correlations used to predict liquid entrainment fraction (E) in gas–liquid annular flow:

$$\left(\frac{dp}{dz}\right)_a = G^2 \frac{d}{dz}\left[\frac{x^2}{\alpha_G \rho_G} + \frac{(1 - x)^2}{(1 - \alpha_G)\rho_L}\right] \tag{5.12}$$

$$\left(\frac{dp}{dz}\right)_a = G^2 \frac{d}{dz}\left[\frac{x^2}{\alpha_G \rho_G} + \frac{(1 - E)^2(1 - x)^2}{\beta \rho_L} + \frac{E^2(1 - x)^2}{(1 - \alpha_G - \beta)\rho_L}\right] \tag{5.13}$$

The volume fraction occupied by the liquid film (β) defined by Eq. (5.14) does not consider the fraction of liquid droplets entrained in the gas core and hence may not be confused with liquid holdup (α_L). However, it is evident that under the condition of $E \approx 0$, the volume fraction of liquid film is equal to the liquid holdup, i.e., $\beta = \alpha_L$. Equation (5.14) is based on a valid assumption that entrained liquid droplets travel at a velocity same as that of the gas:

$$\beta = 1 - \alpha_G - \frac{\alpha_G E(1 - x)}{x(\rho_L/\rho_G)} \tag{5.14}$$

It must be reiterated that this is required only in the case of annular flow with considerable liquid entrainment. For the case of annular flow at low system pressures ($E \approx 0$) and at flow patterns other than annular flow ($E = 0$), hence it is simplified back to Eq. (5.12) such that the accelerational pressure drop at a given location can be determined based on void fraction and mass flux of each phase. For the case of adiabatic two-phase flow over a short pipe length, two-phase flow quality and void fraction can be assumed to remain constant and hence $(dp/dz)_a \approx 0$.

Finally, the frictional component of two-phase pressure drop is calculated as

$$\left(\frac{dp}{dz}\right)_f = \Phi_j^2 \left(\frac{dp}{dz}\right)_j \tag{5.15}$$

The two-phase frictional multiplier Φ^2 can be expressed in several different forms depending upon the assumption of flow of either single phase through the pipe. Broadly, four different cases of two-phase frictional multiplier can be defined as shown in Eqs. (5.16, 5.17, 5.18, and 5.19). In these equations, single-phase pressure drop is the single-phase frictional pressure drop unless otherwise specified, and $(dp/dz)_f$ would always represent frictional component of the two-phase pressure drop:

$$\left(\frac{dp}{dz}\right)_f = \Phi_{LO}^2 \left(\frac{dp}{dz}\right)_{LO} \quad \text{where} \quad \left(\frac{dp}{dz}\right)_{LO} = \frac{2f_{LO}G^2}{D\rho_L} \tag{5.16}$$

$$\left(\frac{dp}{dz}\right)_f = \Phi_L^2 \left(\frac{dp}{dz}\right)_L \quad \text{where} \quad \left(\frac{dp}{dz}\right)_L = \frac{2f_L G^2 (1 - x)^2}{D\rho_L} \tag{5.17}$$

$$\left(\frac{dp}{dz}\right)_f = \Phi_{GO}^2 \left(\frac{dp}{dz}\right)_{GO} \quad \text{where} \quad \left(\frac{dp}{dz}\right)_{GO} = \frac{2f_{GO}G^2}{D\rho_G} \tag{5.18}$$

$$\left(\frac{dp}{dz}\right)_f = \Phi_G^2 \left(\frac{dp}{dz}\right)_G \quad \text{where} \quad \left(\frac{dp}{dz}\right)_G = \frac{2f_G G^2 x^2}{D\rho_G} \tag{5.19}$$

Two-phase flow literature provides several frictional pressure drop correlations based on these different two-phase frictional multipliers (Φ^2); see Eqs. (5.16, 5.17, 5.18, and 5.19). Next, some of the top-performing pressure drop correlations used to determine frictional component of the two-phase pressure drop will be briefly

Table 5.1 Values of constant C to be used in Lockhart and Martinelli (1949) correlation, Eq. (5.20), for different single-phase flow regimes

Liquid phase	Gas phase	C
Turbulent	Turbulent	20
Laminar	Turbulent	12
Turbulent	Laminar	10
Laminar	Laminar	5

discussed. For a more detailed discussion on the separated flow model, refer to Ghajar and Bhagwat (2017).

Lockhart and Martinelli (1949) Correlation They proposed one of the first such correlations. Their correlation expresses two-phase frictional multipliers Φ_L^2 and Φ_G^2 as a function of X parameter and is expressed by Eq. (5.20). The Lockhart and Martinelli (1949) parameter (X) is given by Eq. (5.21) where the single-phase friction factors (f_L and f_G) are found using appropriate single-phase friction factors such as those of Blasius (1913) and Churchill (1977):

$$\Phi_L^2 = 1 + C/X + 1/X^2 \quad \text{or} \quad \Phi_G^2 = 1 + CX + X^2 \tag{5.20}$$

$$X = \left[\frac{(dp/dz)_{f,L}}{(dp/dz)_{f,G}}\right]^{0.5} = \frac{1-x}{x}\left[\frac{f_L}{f_G}\frac{\rho_G}{\rho_L}\right]^{0.5} = \frac{U_{SL}}{U_{SG}}\left[\frac{f_L}{f_G}\frac{\rho_L}{\rho_G}\right]^{0.5} \tag{5.21}$$

The value of C used in Eq. (5.20) given by Chisholm (1967) is given in Table 5.1, and it depends on the laminar or turbulent nature of the gas and liquid phase, respectively. A worked-out example problem illustrating the use of Lockhart and Martinelli (1949) correlation is provided at the end of Chap. 7.

Muller-Steinhagen and Heck (1986) Correlation They developed a correlation to extrapolate between single-phase liquid and single-phase gas flow. Their correlation can be used for both adiabatic and evaporating two-phase flow conditions (provided the two-phase flow quality distribution across the pipe length is known). The physical form of their correlation is as shown in Eq. (5.22). The parameter Y defined in Eq. (5.23) is the same as that given by Chisholm (1973). The correlation is verified against 9300 measurements including data of air–oil, air–water, and steam–water fluid combinations and pipe diameters ranging from 4 to 39 mm I.D. The application of Eq. (5.22) is restricted to $Re_{LO} > 100$ and $Y^2 > 1$. Note that for determination of single-phase friction factor f_j, their correlation separates single-phase laminar and turbulent regions using a threshold value of $Re_j = 1187$ where "j" can be gas or liquid phase. Muller-Steinhagen and Heck (1986) claimed their correlation to predict the two-phase frictional pressure drop data with an average error of 41.7% with only 49.5% of the data points predicted within ±30% error bands:

$$\Phi_{LO}^2 = (1 - x)^{0.33}\left(1 + 2x\left(Y^2 - 1\right)\right) + Y^2 x^3 \tag{5.22}$$

$$Y = \sqrt{\frac{(dp/dz)_{GO}}{(dp/dz)_{LO}}} \tag{5.23}$$

Xu and Fang (2012) Correlation They proposed a two-phase frictional pressure drop correlation for evaporating two-phase flow of refrigerants in horizontal pipe. Their correlation is based on 2622 data points of 14 refrigerants (R11, R12, R22, R134a, R32, R407C, R507, R507A, R410A, CO_2, R404A, R32/R125, R123, and ammonia) for $1 < D_h < 19$ mm (circular and rectangular pipe geometries) and $25 < G < 1150$ kg/m^{2}s. They found that the pipe hydraulic diameter and gas–liquid interface surface tension affect the two-phase frictional pressure drop significantly than any other two-phase flow parameter, and hence their correlation accounts for these two variables through the inclusion of non-dimensional Laplace number. The physical form of their correlation is as expressed in Eq. (5.24):

$$\Phi_{LO}^2 = \left[Y^2 x^3 + (1 - x)^{0.33}\left(1 + 2x\left(Y^2 - 1\right)\right)\right]\left(1 + 1.54(1 - x)^{0.5}\mathrm{La}\right) \tag{5.24}$$

In this equation Y is calculated from Eq. (5.23), while La is the Laplace number defined by Eq. (4.19). Their correlation is claimed to predict the two-phase frictional pressure drop data with a mean absolute relative deviation of 25.2%. It should be noted that Xu and Fang (2012) recommend use of Fang et al. (2011) correlation to calculate single-phase fanning friction factors f_{LO} and f_{GO}. The correlations of Fang et al. (2011) for single-phase liquid and gas fanning friction factors are reported in Eqs. (5.25a) and (5.25b):

$$f_{LO} = 0.0625\left[\log\left(\frac{150.39}{\mathrm{Re}_{LO}^{0.98865}} - \frac{152.66}{\mathrm{Re}_{LO}}\right)\right]^{-2} \tag{5.25a}$$

$$f_{GO} = 0.0625\left[\log\left(\frac{150.39}{\mathrm{Re}_{GO}^{0.98865}} - \frac{152.66}{\mathrm{Re}_{GO}}\right)\right]^{-2} \tag{5.25b}$$

Zhang et al. (2010) Correlation For mini/micro channels having pipe diameters in a range of 0.1–6 mm, Zhang et al. (2010) proposed a two-phase frictional pressure drop correlation (for adiabatic two-phase flow) based on the experimental data of two-phase flow of refrigerants, air–water and air–ethanol. Their correlation for two-phase frictional multiplier is the same as Eq. (5.20) proposed by Lockhart and Martinelli (1949), however with the parameter C as a variable given by Eq. (5.26):

$$C = \begin{pmatrix} 21[1 - \exp{(-0.674/\text{La})}] & : (\text{two} - \text{component two} - \text{phase flow}) \\ 21[1 - \exp{(-0.142/\text{La})}] & : (\text{one} - \text{component two} - \text{phase flow}) \end{pmatrix}$$
$$(5.26)$$

Bhagwat (2015) Correlation Ghajar and Bhagwat (2014b) and later on Bhagwat (2015) did a comprehensive study of 16 top-performing two-phase frictional pressure drop correlations based on separated flow model available in the two-phase flow literature. The performance of these correlations was assessed using a consolidated database of 9623 data points (horizontal and upward inclined two-phase flow) amassed from 60 data sources in the two-phase flow literature and consisted of air–water, air–oil, refrigerants (R12, R22, R123, R134a, R236fa, R245fa, R404A, R410A, R507, R32+R125, R290 (propane), R1234yf, R422D, R717 (ammonia), R744 (CO_2)), and other miscellaneous fluid combinations such as air–glycerin and air–kerosene. The analysis of the two-phase frictional pressure drop correlations against the experimental data bank showed that the frictional pressure drop correlation of Muller-Steinhagen and Heck (1986) given by Eq. (5.22) performed relatively better than the other correlations; see Bhagwat (2015) for additional details. As mentioned earlier, the correlation of Muller-Steinhagen and Heck (1986) is restricted to $\text{Re}_{\text{LO}} > 100$ and $Y^2 > 1$ and does not have the flexibility of accommodating a wide range of fluid properties, flow rates, pipe diameters, and pipe orientations. To expand on the performance range of correlation of Muller-Steinhagen and Heck (1986), Bhagwat (2015), using the consolidated database of 9623 data points explained earlier, modified the Muller-Steinhagen and Heck (1986) correlation in the form of Eq. (5.27) by introducing three new variables B_1, B_2, and B_3 in Eq. (5.22) given by Eqs. (5.28, 5.29, and 5.30). Note that in Eq. (5.27) by setting $B_1 = 2$, $B_2 = 1$, and $B_3 = 0$, the original Muller-Steinhagen and Heck (1986) correlation given by Eq. (5.22) is obtained. Recall that the parameter Y used in Eq. (5.27) was defined in Eq. (5.23):

$$\Phi_{\text{LO}}^2 = \left(\left[(1 - x)^{0.33} \left(1 + B_1 x \left(Y^2 - 1 \right) \right) \right] + B_2 Y^2 x^3 \right) \left(1 + B_3 \left(1 - x \right)^2 \right) \quad (5.27)$$

$$B_1 = [0.85 + 1.703(1 - \exp{(-6.25 \zeta \times \text{Bo})})] \times \Pi_1 \Pi_2 \Pi_3 \quad (5.28)$$

$$B_2 = 1 - \sqrt{\rho_G / \rho_L} \quad (5.29)$$

$$B_3 = \begin{cases} -0.3(1 + \sin\theta)^{-16.25} + 0.3 & : 0° \le \theta \le +20° \\ -0.012(1 + \sin\theta)^{4.1} + 0.34 & : +20° < \theta \le +90° \end{cases} \tag{5.30}$$

$$\Pi_1 = \left(1 + 2.65\left[1 - \exp\left(-1.677 N_{\mu L}\right)\right]\right) \tag{5.31}$$

$$\Pi_2 = \begin{cases} 0.55 & : \zeta \le 1.0, \ \text{Bo} \ge 1 \\ 1 & : \zeta > 1.0, \ \text{Bo} < 1 \end{cases} \tag{5.32}$$

$$\Pi_3 = \left[1 + 0.005\left(\frac{1-x}{x}\right)\right]^{0.5} \tag{5.33}$$

$$\zeta = 2.5\sqrt{\frac{\rho_L}{\rho_{ref}}}\left(\frac{\mu_G}{\mu_L}\right)^{0.25} \tag{5.34}$$

$$\text{Bo} = \frac{g(\rho_L - \rho_G)(D_h/2)^2}{\sigma} \tag{5.35}$$

$$N_{\mu L} = \frac{\mu_L}{\left[\rho_L \sigma \sqrt{\frac{\sigma}{g(\rho_L - \rho_G)}}\right]^{0.5}} \tag{5.36}$$

Referring to Eq. (5.27), the parameters B_1 (multiplying factor to the term $Y^2 - 1$) given by Eq. (5.28) and B_2 (multiplying factor to the term Y^2) given by Eq. (5.29) control the slope of Φ_{LO}^2 vs. x for lower and higher values of quality, respectively, whereas parameter B_3 given by Eq. (5.30) simply acts as a multiplier to Φ_{LO}^2 that modifies the overall value of two-phase pressure drop as a function of pipe orientation. As expressed by Eq. (5.28), the parameter B_1 is modeled as a function of pipe hydraulic diameter (D_h), fluid properties (ρ_L, ρ_G, μ_L, and σ), and the two-phase flow quality (x). The effect of these two-phase flow variables on B_1 and hence Φ_{LO}^2 is accounted in the form of non-dimensional numbers such as Bond number (Bo) and viscosity number ($N_{\mu L}$) given by Eqs. (5.35) and (5.36), respectively. The variables Π_1, Π_2, and Π_3 used in parameter B_1 are expressed by Eqs. (5.31), (5.32), and (5.33), respectively. Parameter B_2 given by Eq. (5.29) is modeled as a function of the density ratio (ρ_G/ρ_L) and is found to vary in a range of 0.5–1.0. As shown in Eq. (5.30), parameter B_3 is modeled as a function of pipe orientation such that its value varies in a range of 0–0.3. Based on the experimental data used, the two different equations (of the form $y = ax^b + c$) are chosen for near horizontal and steeper pipe inclinations in the upward direction. Note that there is a good agreement (within 1%) between the extrapolated values of B_3 for two cases of $0° \leq \theta \leq +20°$ ($B_3 = 0.299$) and $+20° < \theta \leq +90°$ ($B_3 = 0.297$). The term $(1 - x)^2$ used in conjunction of B_3 ensures that the predicted Φ_{LO}^2 is weighted such that it gradually decreases with increase in the two-phase flow quality or alternatively as the flow patterns become inertia driven in nature and hence relatively insensitive to the change in the pipe orientation.

Getting back to parameter B_1, let us look at its physical significance and need for including parameters Π_1, Π_2, and Π_3. The parameter B_1 is modeled as a variable such that it is sensitive to the increase in Bond number in the vicinity of Bo ≈ 1 ($0.7 < $ Bo < 1.3) and thereafter follows a saturation trend. Note that this threshold value of Bo after which saturated trend ensues is subject to a small change with variation in property group ζ given by Eq. (5.34). The inclusion of Bond number accounts for the balance between body and surface tension forces on the two-phase frictional multiplier. For small diameter pipes, effect of surface tension on two-phase flow is more significant compared to the effect of gravity. Accordingly, for Bo < 1 the surface tension effects are prominent in comparison to the body forces, whereas for Bo > 1 the effect of surface tension on the two-phase flow is inconsequential.

The data bank used by Bhagwat (2015) showed that the condition of Bo < 1 corresponds to pipe diameters in a range of 0.069–4.5 mm and consists of refrigerants (1822 data points) as well as air–water (645 data points) fluid combinations. According to the observations of Kandlikar (2002), approximately this range of pipe diameter ($0 < D \leq 3$ mm) represents the micro to mini scale two-phase flow phenomenon. The variable B_1 as a function of Bo increases the value of Φ_{LO}^2 with increase in the pipe diameter (due to increase in the gas–liquid interfacial area).

Variable B_1 also accounts for the effect of liquid dynamic viscosity through parameter Π_1 which is modeled as a function of viscosity number ($N_{\mu L}$); see Eq. (5.36). Increase in liquid phase viscosity increases the shear in the liquid phase and hence increases the two-phase frictional multiplier and consequently the two-phase pressure drop.

The parameter Π_2 switches between a constant value of 0.55 and 1.0 for two different cases based on Bo and ζ as shown in Eq. (5.32). According to Bhagwat (2015), analysis of the experimental data used showed that in comparison to liquid refrigerant and its vapor, data for air–water, air–oil, air–kerosene, and air–glycerin fluid combinations is overpredicted (parallel shift) while using Eq. (5.28). To reduce the magnitude of this overprediction, use of multiplying factor less than unity (0.55 in this case) was found to give the best fit. Using the property group of ζ defined by Eq. (5.34), these fluid combinations could be separated from the refrigerant data. Note that ρ_{ref} used in definition of ζ is reference value of the density equal to the density of liquid water taken as 1000 kg/m³.

Finally, the variable Π_3 as a function of two-phase flow quality (x) is modeled such that for small values of x, $\Pi_3 > 1$, whereas for $x \to 1$, $\Pi_3 \approx 1$; see Eq. (5.33).

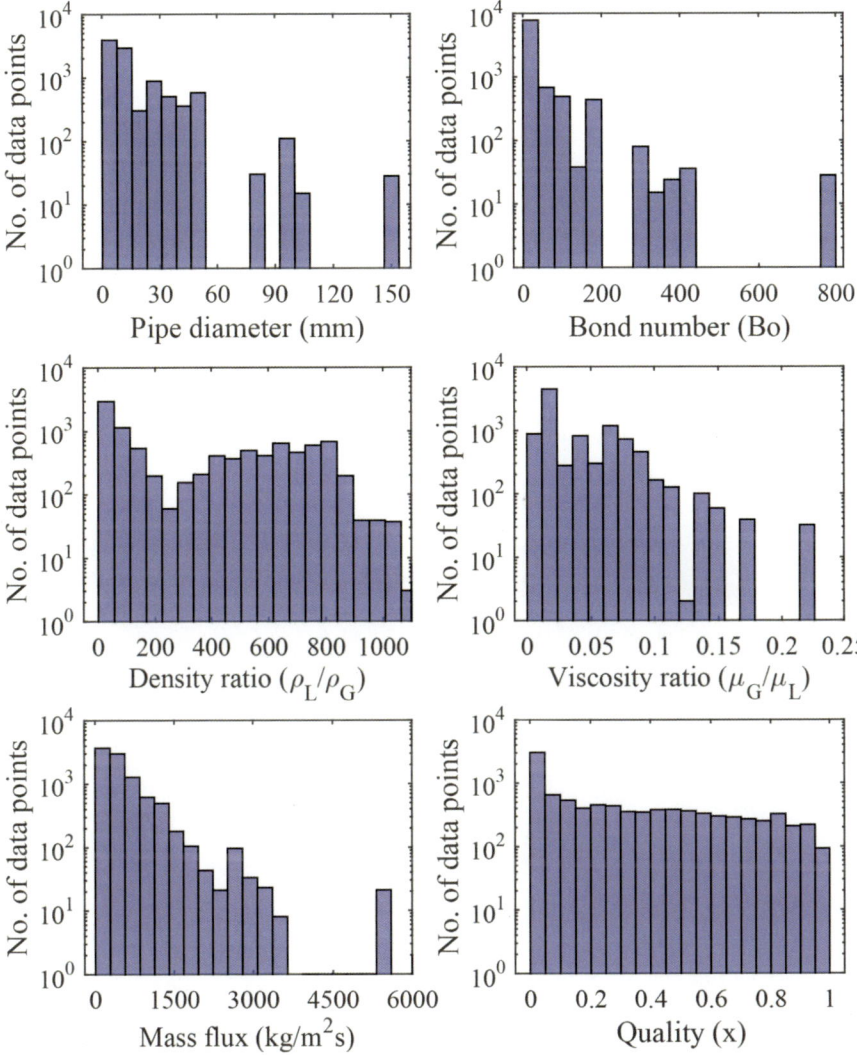

Fig. 5.10 Distribution of experimental data bank (9623 data points) used by Bhagwat (2015)

The small values of x typically correspond to bubbly and slug flow patterns with liquid and gas phase in laminar–laminar or turbulent–laminar flow regimes. Thus, the predicted values of two-phase pressure drop are sensitive to Π_3 only in these flow regimes without affecting the predicted data in laminar–turbulent and turbulent–turbulent flow regimes.

The distribution of experimental data at varying flow conditions used for the development of Bhagwat (2015) correlation is shown in Fig. 5.10. The distribution of hydraulic pipe diameter is such that it occupies micro to macro scale two-phase flow ($0.069 \leq D_\mathrm{h} \leq 152$ mm) that corresponds to a wide range of Bond numbers

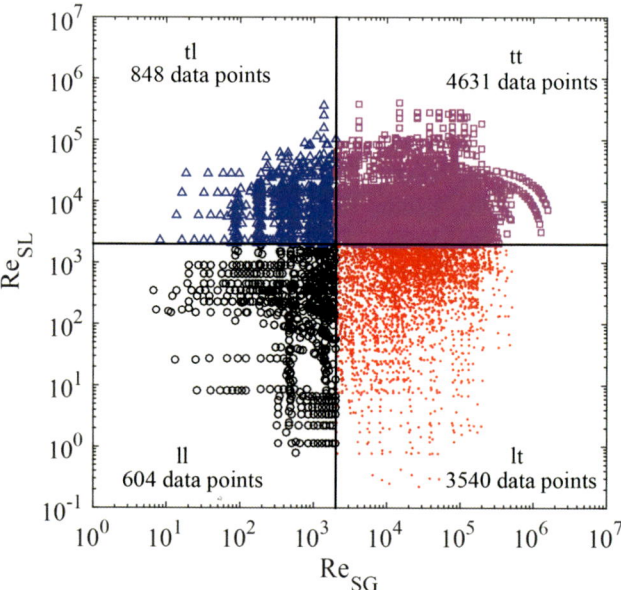

Fig. 5.11 Distribution of experimental data in different flow regimes (ll laminar–laminar, lt laminar–turbulent, tl turbulent–laminar, tt turbulent–turbulent)

(0.0015 to 800). It is seen that the two-phase flow quality is fairly distributed over the entire range ($0 < x < 1$) and hence is expected to occupy the entire range of flow patterns typically observed in gas–liquid two-phase flow. The experimental data is also observed to be distributed over a wide range of two-phase mixture mass flux. The experimental database also consists of a wide range of density ratio (ρ_L/ρ_G). The low values of density ratio correspond to high system pressure data such as that for refrigerants (typically for CO_2, propane, ammonia, and refrigerants at higher saturation temperature), while high-density ratio data consist of air–water and air–glycerin fluid combinations. In addition to the density ratio, a wide range of viscosity ratio (μ_G/μ_L) is also observed. The lower and higher values of (μ_G/μ_L) typically correspond to the air–oil and refrigerants two-phase flow data, respectively. Additionally, the distribution of experimental data for four different cases of laminar–laminar (ll), laminar–turbulent (lt), turbulent–laminar (tl), and turbulent–turbulent (tt) two-phase flow is shown in Fig. 5.11. From experience it can be said that the laminar–laminar and turbulent–laminar flow regions typically consist of slug/plug and bubbly/bubbly–slug flow patterns, respectively, whereas the laminar–turbulent and turbulent–turbulent flow regions may consist of wavy slug (intermittent) and annular flow patterns. A brief summary of the range of different two-phase flow parameters used in the development of Bhagwat (2015) correlation is reported in Table 5.2.

The performance of the Bhagwat (2015) correlation is evaluated based on the statistical parameters such as mean relative deviation (MRD) and mean absolute relative deviation (MARD). Note that achieving a low value of MRD or MARD is

Table 5.2 Range of experimental data used for development of Bhagwat (2015) two-phase frictional pressure drop correlation

Parameter	Range
Hydraulic pipe diameter (mm)	0.1–152
Mixture mass flux (kg/m^{2}s)	5–7000
Two-phase flow quality	0.0001–0.99
Liquid to gas density ratio (ρ_L/ρ_G)	5–920
Liquid to gas viscosity ratio (μ_L/μ_G)	10–5000
Surface tension (N/m)	0.002–0.075
Pipe orientation (θ)	$0° \leq \theta \leq +90°$
Bond number (Bo)	0.0015–800
Liquid only Reynolds number (Re$_{LO}$)	$1 - 2.85 \times 10^5$
Gas only Reynolds number (Re$_{GO}$)	$1800 - 1.5 \times 10^7$
Viscosity number ($N_{\mu L}$)	0.0008–4
Lockhart-Martinelli parameter (X)	0.002–800
Gas volumetric flow fraction (λ)	0.005–0.99
Flow patterns	All except stratified

not the absolute measure to gauge the accuracy of any correlation since these parameters (MRD and MARD) are strongly influenced by the number of data points and distribution of the error. It is also important for the correlation to be able to predict maximum number of data points within certain error bands. Accordingly, the accuracy of the correlation is also compared based on the percentage of data points predicted within ±30% (N_{30}) and ±50% (N_{50}) error bands. Considering the practical difficulty in modeling of two-phase pressure drop (as a function of several two-phase flow variables), use of higher error bands up to ±50% in the prediction of two-phase pressure drop is considered as acceptable and has become a norm in the two-phase flow literature.

The goodness of the correlation of Bhagwat (2015) for different cases of single-phase flow regimes (laminar/turbulent) is shown in Fig. 5.12. The figure shows the prediction of correlation of Bhagwat (2015) against the measured data for different combinations of single-phase flow regimes. For laminar–laminar (Re$_{SL} \leq 2000$, Re$_{SG} \leq 2000$), laminar–turbulent (Re$_{SL} \leq 2000$, Re$_{SG} > 2000$), and turbulent–turbulent (Re$_{SL} > 2000$, Re$_{SG} > 2000$) flow regimes, the correlation of Bhagwat (2015) successfully predicts more than 70% and 90% of data points within ±30% and ±50% error bands, respectively. For the laminar–turbulent and turbulent–turbulent cases, MRD is less than 4%, and MARD is 25% or less. Accuracy of the correlation of Bhagwat (2015) slightly drops down for the case of turbulent–laminar flow (Re$_{SL} > 2000$, Re$_{SG} \leq 2000$). A careful observation of the experimental data shows that a majority of data in this flow regime belongs to the slug/churn (intermittent flow) transition in upward inclined two-phase flow where the pressure drop increases suddenly due to slug breakup resulting into vigorous mixing and chaotic two-phase flow behavior. Consequently, all two-phase flow models underpredict the frictional pressure drop for this case.

Figure 5.13 shows the performance of Muller-Steinhagen and Heck (1986) correlation for different combinations of single-phase flow regimes. Similar to the

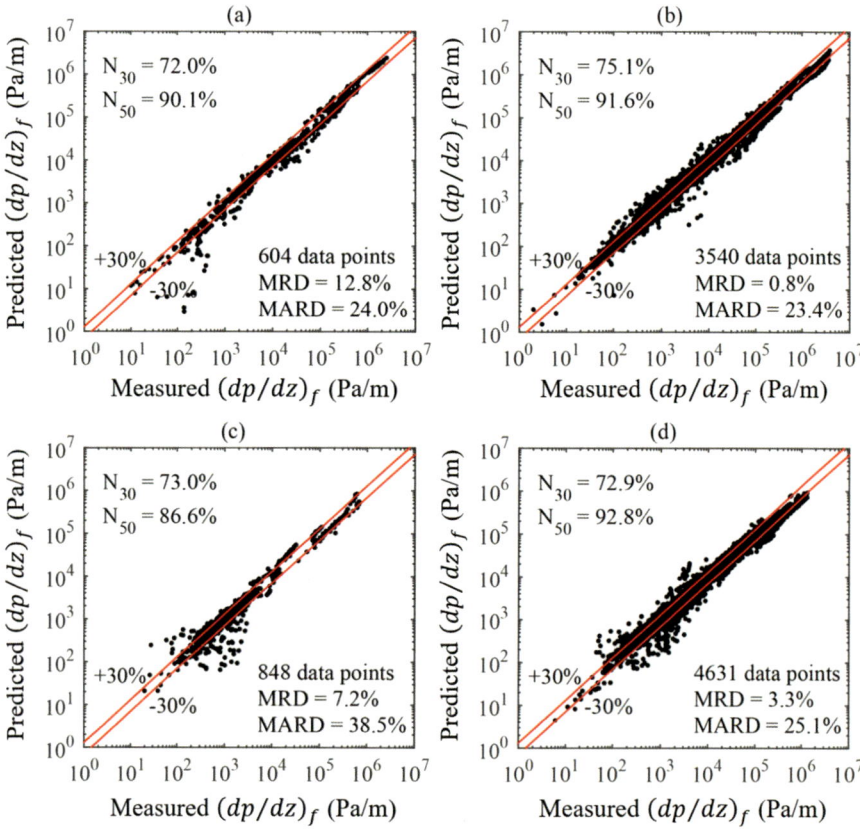

Fig. 5.12 Prediction of Bhagwat (2015) correlation against the entire data for different single-phase flow regimes (all data, 9623 data points) (**a**) Laminar-Laminar, (**b**) Laminar-Turbulent, (**c**) Turbulent-Laminar, (**d**) Turbulent-Turbulent

Bhagwat (2015) correlation, accuracy of Muller-Steinhagen and Heck (1986) is found to deteriorate for the case of turbulent–laminar flow. Considering the overall accuracy in terms of N_{30}, N_{50}, MRD, and MARD, the Bhagwat (2015) correlation performs consistently better than Muller-Steinhagen and Heck (1986) correlation for all four cases of single-phase flow regimes. Finally, the performance of the correlation of Bhagwat (2015) over the entire data (9623 data points) consisting of horizontal, vertical, and upward pipe inclinations is presented graphically in Fig. 5.14. For the consolidated data bank, correlation of Bhagwat (2015) predicts 74% and 92% of data within $\pm30\%$ and $\pm50\%$ error bands with a MRD of 3.3% and MARD of 25.6%. In comparison to this, as shown by Fig. 5.15, Muller-Steinhagen and Heck(1986) correlation predicts 63% and 88% of data points within $\pm30\%$ and $\pm50\%$ error bands with a MRD of 8.0% and MARD of 28.8%.

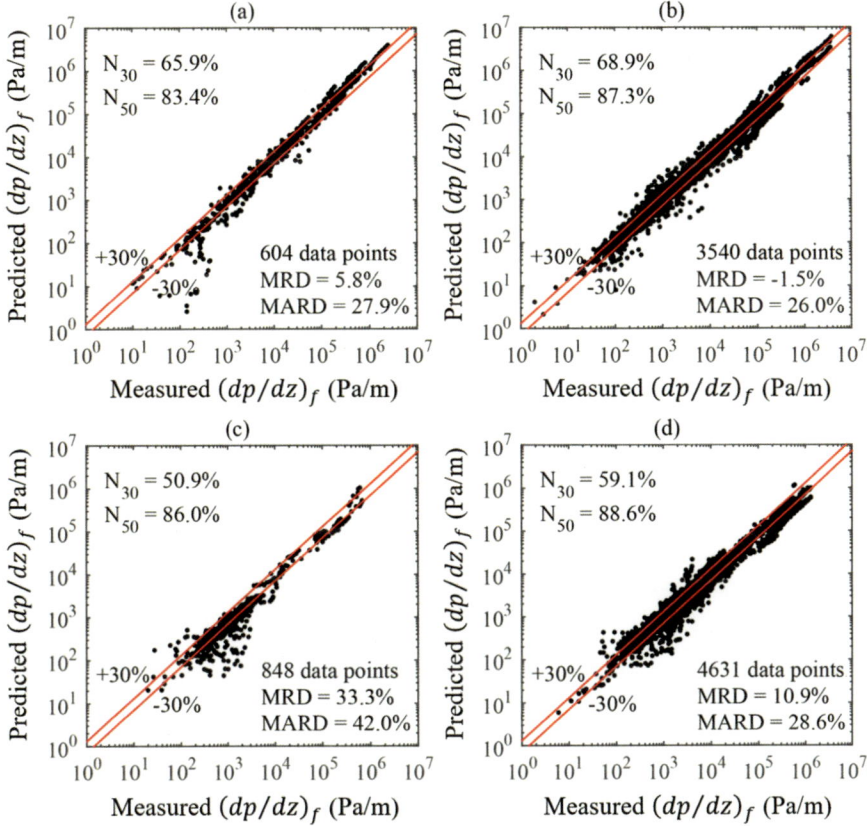

Fig. 5.13 Prediction of Muller-Steinhagen and Heck (1986) correlation against the entire data for different single-phase flow regimes (all data, 9623 data points) (**a**) Laminar-Laminar, (**b**) Laminar-Turbulent, (**c**) Turbulent-Laminar, (**d**) Turbulent-Turbulent

An Illustrative Example: Use of Xu and Fang (2012) Two-Phase Frictional Pressure Drop Correlation Consider a boiling two-phase flow of R134a through a 6 mm I.D. horizontal copper tube at a mass flux of 800 kg/m²s and a system pressure of 1500 kPa. The two-phase refrigerant enters the 1 m long tube with a quality of $x = 0.3$ and exits with a quality of $x = 0.7$. Calculate the two-phase frictional pressure drop for this flow using the two-phase frictional pressure drop correlation of Xu and Fang (2012). For simplicity, assume that the quality changes linearly with the pipe length, and use a mean quality of 0.5.

For refrigerant R134a at 1500 kPa, use the following thermophysical properties:

Liquid density $(\rho_L) = 1078$ kg/m³, gas density $(\rho_G) = 76.95$ kg/m³, liquid viscosity $(\mu_L) = 0.0001746$ Pa.s, gas viscosity $(\mu_G) = 0.0000138$ Pa.s, and surface tension $(\sigma) = 0.00427$ N/m.

Fig. 5.14 Performance of Bhagwat (2015) correlation against entire data (9623 data points)

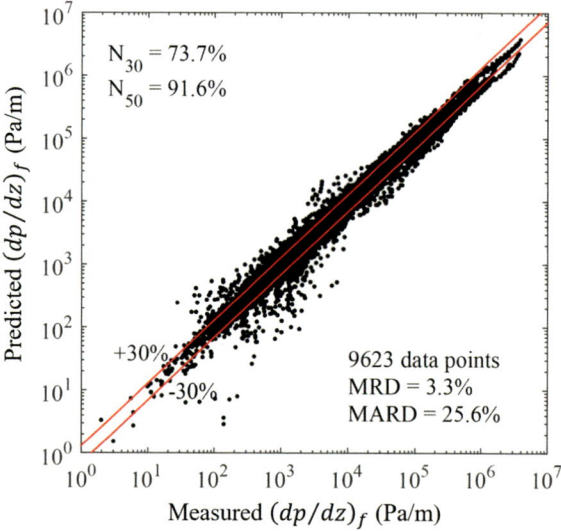

Fig. 5.15 Performance of Muller-Steinhagen and Heck (1986) correlation against entire data (9623 data points)

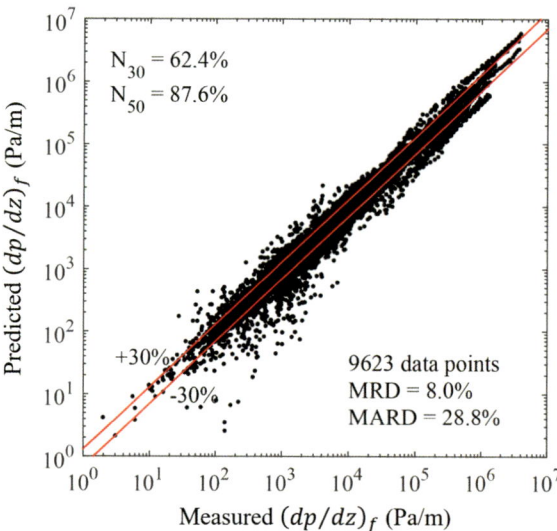

Solution

The two-phase frictional pressure drop for Xu and Fang (2012) is determined from Eq. (5.24):

$$\Phi_{LO}^2 = \left[Y^2 x^3 + (1 - x)^{0.33} \left(1 + 2x(Y^2 - 1) \right) \right] \left(1 + 1.54(1 - x)^{0.5} \text{La} \right)$$

where from Eq. (5.23) $Y = \sqrt{\dfrac{(dp/dz)_{GO}}{(dp/dz)_{LO}}}$ and from Eq. (4.19) $\text{La} = \sqrt{\dfrac{\sigma}{g(\rho_L - \rho_G)}} \dfrac{1}{D_h}$

It should be noted that Xu and Fang (2012) recommend use of Fang et al. (2011) correlations to calculate the single-phase fanning friction factors f_{LO} and f_{GO}. The Fang et al. (2011) correlations given by Eqs. (5.25a) and (5.25b) are

$$f_{LO} = 0.0625 \left[\log \left(\frac{150.39}{Re_{LO}^{0.98865}} - \frac{152.66}{Re_{LO}} \right) \right]^{-2}$$

$$f_{GO} = 0.0625 \left[\log \left(\frac{150.39}{Re_{GO}^{0.98865}} - \frac{152.66}{Re_{GO}} \right) \right]^{-2}$$

The liquid and gas Reynolds numbers (Re_{LO} and Re_{GO}) required in Eqs. (5.25a) and (5.25b) are calculated by the following equations:

$$Re_{LO} = \frac{GD}{\mu_L} = \frac{800 \times 0.006}{0.0001746} = 27491$$

$$Re_{GO} = \frac{GD}{\mu_G} = \frac{800 \times 0.006}{0.0000138} = 347826$$

Thus from the friction factor correlations of Fang et al. (2011), Eqs. (5.25a) and (5.25b), $f_{LO} = 0.005995$ and $f_{GO} = 0.0035175$.

The single-phase pressure drop for liquid and gas phases are calculated from Eqs. (5.16) and (5.18):

$$\left(\frac{dp}{dz} \right)_{LO} = \frac{2f_{LO}G^2}{D\rho_L} = \frac{2 \times 0.005995 \times 800^2}{0.006 \times 1078} = 1186 \text{ Pa/m}$$

$$\left(\frac{dp}{dz} \right)_{GO} = \frac{2f_{GO}G^2}{D\rho_G} = \frac{2 \times 0.0035175 \times 800^2}{0.006 \times 76.95} = 9752 \text{ Pa/m}$$

Thus, from Eq. (5.23), $Y = \sqrt{\frac{(dp/dz)_{GO}}{(dp/dz)_{LO}}} = \sqrt{\frac{9752}{1186}} = 2.867$

And from Eq. (4.19), $La = \sqrt{\frac{\sigma}{g(\rho_L - \rho_G)}} \frac{1}{D_h} = \sqrt{\frac{0.00427}{9.81(1078 - 76.95)}} \frac{1}{0.006} = 0.1099$

The two-phase frictional multiplier (Φ_{LO}^2) is calculated from Eq. (5.24) to be

$$\Phi_{LO}^2 = \left[Y^2 x^3 + (1 - x)^{0.33} \left(1 + 2x(Y^2 - 1) \right) \right] \left(1 + 1.54(1 - x)^{0.5} La \right)$$

$$= \left[2.867^2 \times 0.5^3 + (1 - 0.5)^{0.33} (1 + 2 \times 0.5(2.867^2 - 1)) \right]$$

$$\times \left(1 + 1.54(1 - 0.5)^{0.5} \times 0.1099 \right)$$

$$= 8.455$$

Finally, the two-phase frictional pressure drop is determined from Eq. (5.16) as

$$\left(\frac{dp}{dz} \right)_f = \Phi_{LO}^2 \left(\frac{dp}{dz} \right)_{LO} = 8.455 \times 1186 = 10{,}028 \ \mathrm{Pa/m}$$

Discussion The total pressure drop in gas–liquid two-phase flow is the sum of hydrostatic, frictional, and accelerational components as shown in Eq. (2.1). For the horizontal pipe orientation of this problem, the hydrostatic pressure drop is zero, and the frictional pressure drop was calculated to be 10,028 Pa/m. The remaining accelerational component of two-phase pressure drop is due to the expansion of gas phase as the two-phase mixture travels downstream and is trivial for non-boiling two-phase flow and hence can be neglected. However, in the case of boiling two-phase flow (problem under consideration), the pressure drop due to acceleration of the gas phase can offer a significant contribution to the total pressure drop depending upon the pipe diameter and pipe orientation. It should be noted that similar to hydrostatic component of pressure drop (see Eq. 2.2), the calculation of accelerational component of two-phase pressure drop also requires correct estimation of void fraction. The accelerational two-phase pressure drop component is expressed by Eq. (5.37) as given by Ghajar and Bhagwat (2014b):

$$\left(\frac{dp}{dz} \right)_a = \frac{1}{L}$$

$$\times \left\{ \left[\frac{G_L^2}{\rho_L \ (1 - \alpha_G)} - \frac{G_G^2}{\rho_G \ \alpha_G} \right]_{out} - \left[\frac{G_L^2}{\rho_L \ (1 - \alpha_G)} - \frac{G_G^2}{\rho_G \ \alpha_G} \right]_{in} \right\}$$

$$\tag{5.37}$$

The void fraction at the inlet and outlet of the pipe needed in Eq. (5.37) for the calculation of the accelerational component of two-phase pressure drop was determined from the void fraction correlation of Woldesemayat and Ghajar (2007) to be $\alpha_{G, \ in} = 0.783$ and $\alpha_{G, \ in} = 0.926$. The final value of accelerational pressure drop calculated from Eq. (5.37) is 4064 Pa/m. Using Eq. (2.1), the total pressure drop for this problem is

$$\left(\frac{dp}{dz}\right)_t = \left(\frac{dp}{dz}\right)_h + \left(\frac{dp}{dz}\right)_f + \left(\frac{dp}{dz}\right)_a = 0 + 10,028 + 4064 = 14,092 \text{ Pa/m}$$

Therefore, the accelerational pressure drop contributes to about 29% of the total two-phase pressure drop, while the remaining 71% contribution is due to the frictional two-phase pressure drop. It should be noted that in the case of adiabatic two-phase flow, the quality and void fraction at pipe inlet and exit remain practically unchanged, and hence the accelerational two-phase pressure drop may be neglected.

Chapter 6
Entrainment

Background

A peculiar phenomenon known as entrainment is observed in annular flow due to the relative motion between the gas an liquid phase. The entrainment process is characterized by the flow of tiny liquid droplets into the central fast-moving gas core. The liquid entrainment is a consequence of the significant shear at the gas–liquid interface that causes tearing of the liquid wave crests in form of ligaments and depends on the phase flow rates, pipe diameter and orientation, and the surface tension at the gas–liquid interface. The liquid entrainment fraction (E) is defined as the ratio of mass flow rate/flux of liquid drops entering into the gas core to the total mass flow rate/flux of the liquid phase. Correct knowledge of the rate of liquid entrainment or alternatively the fraction of liquid entrainment is crucial in analyzing heat and mass transfer processes in annular flows. In the case of boiling two-phase flows, accurate estimation of the liquid entrainment fraction is necessary in determination of critical heat flux and dryout conditions.

Entrainment Mechanisms

Ishii and Grolmes (1975) studied the different possible mechanisms responsible for liquid entrainment process. They found that liquid entrainment can happen due to any or all of the wave undercut, wave rolling, wave coalescence, and ripple shearing mechanisms. Additionally, the phenomenon of "bubble burst" due to rupture of liquid-phase crests and "droplet impingement" due to rolling waves may also contribute to the entrainment process.

The different entrainment mechanisms suggested by Ishii and Grolmes (1975) are illustrated in Fig. 6.1. As shown in Fig. 6.2, critical values of gas and liquid flow

© The Author(s), under exclusive license to Springer Nature Switzerland AG 2020
A. J. Ghajar, *Two-Phase Gas-Liquid Flow in Pipes with Different Orientations*,
SpringerBriefs in Applied Sciences and Technology,
https://doi.org/10.1007/978-3-030-41626-3_6

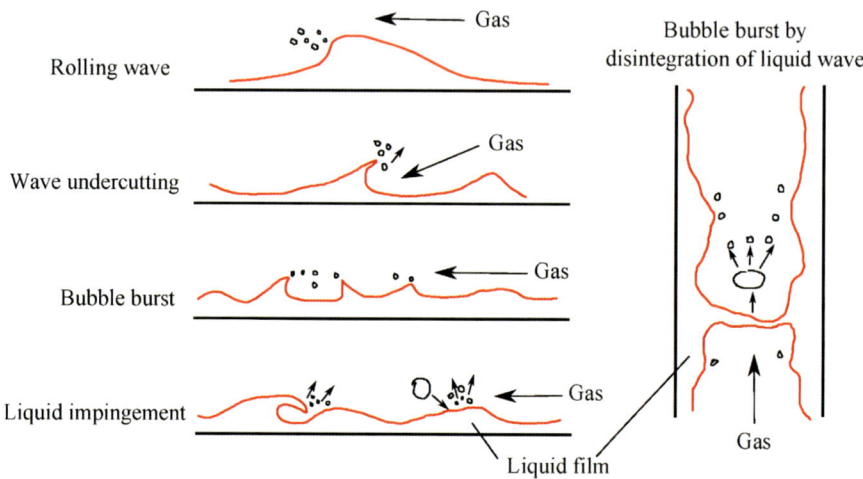

Fig. 6.1 Entrainment mechanisms suggested by Ishii and Grolmes (1975)

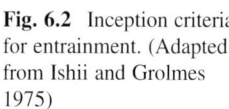

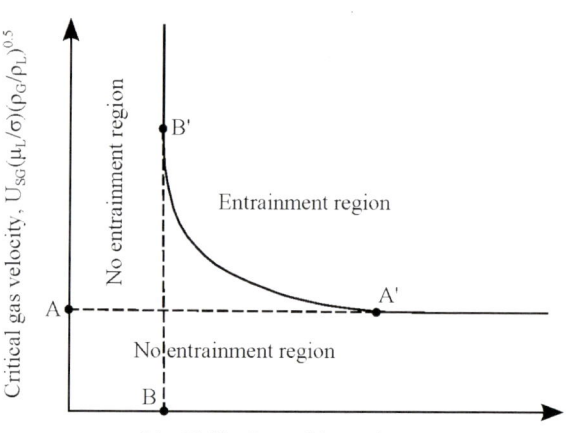

Fig. 6.2 Inception criteria for entrainment. (Adapted from Ishii and Grolmes 1975)

rates are associated with the entrainment process and below which no liquid entrainment is expected to occur. Based on the experimental data, two-phase flow literature offers several correlations to determine the non-dimensional critical gas velocity and the liquid film Reynolds number (Re_{LF}) below which the entrainment phenomenon is negligible or alternatively above which the amount of liquid entrainment is significant (see Fig. 6.2). Note that the Reynolds number based on liquid flowing in form of film is defined by Ishii and Grolmes (1975) as $Re_{LF} = (4\delta U_L \rho_L)/\mu_L$.

One such criterion in terms of liquid film Reynolds number proposed by Ishii and Grolmes (1975) is represented by Eq. (6.1). This equation sets the absolute limit for inception of entrainment process and represents point B in Fig. 6.2 below which entrainment cannot take place. Ishii and Grolmes (1975) found that at point A', $Re_{LF} = 1635$ and beyond which the entrainment affected region is insensitive to the increase in liquid flow rates. The critical gas velocity for Re_{LF} greater than that by Eq. (6.1) is obtained from Eq. (6.2). The critical gas velocity to the left-hand side of this equation is essentially a non-dimensional superficial gas velocity normalized using liquid-phase dynamic viscosity and the surface tension. Note that for Re_{LF} greater than $\text{Re}_{\text{LF, C}}$ but less than 160, critical gas velocity is independent of liquid viscosity number given by Eq. (6.3):

$$\text{Re}_{\text{LF,C}} = 154.7 \left(\frac{\rho_{\text{L}}}{\rho_{\text{G}}} \right)^{0.75} \left(\frac{\mu_{\text{G}}}{\mu_{\text{L}}} \right)^{1.5} \tag{6.1}$$

$$\frac{U_{\text{SG}}\mu_{\text{L}}}{\sigma} \sqrt{\frac{\rho_{\text{G}}}{\rho_{\text{L}}}} \geq \begin{cases} 1.5\text{Re}_{\text{LF}}^{-0.5} & : \text{Re}_{\text{LF,C}} < \text{Re}_{\text{LF}} < 160 \\ 11.78 N_{\mu}^{0.8}\text{Re}_{\text{LF}}^{-0.33} & : N_{\mu} \leq 0.066,\ 160 \leq \text{Re}_{\text{LF}} \leq 1635 \\ 1.35\text{Re}_{\text{LF}}^{-0.33} & : N_{\mu} > 0.066,\ 160 \leq \text{Re}_{\text{LF}} \leq 1635 \\ N_{\mu}^{0.8} & : N_{\mu} \leq 0.066,\ \text{Re}_{\text{LF}} > 1635 \\ 0.1146 & : N_{\mu} > 0.066,\ \text{Re}_{\text{LF}} > 1635 \end{cases} \tag{6.2}$$

$$N_{\mu} = \frac{\mu_{\text{L}}}{\left(\rho_{\text{L}}\sigma\sqrt{\sigma/(g\Delta\rho)} \right)^{0.5}} \tag{6.3}$$

Liquid Entrainment Fraction Correlations

Note that Eqs. (6.1) and (6.2) predict the onset of entrainment and not the magnitude of liquid entrainment fraction. To predict the liquid entrainment fraction, we need specific correlations.

Ishii and Mishima (1989) Correlation Their proposed correlation is applicable to air–water equilibrium annular flow. As shown in Eq. (6.4), their correlation is a function of modified Weber number given by Eq. (6.5). Although the correlation of Ishii and Mishima (1989) consists of non-dimensional numbers as a function of fluid physical properties, it is primarily based on air–water data, and its validity for other fluid combinations must be scrutinized before use:

$$E = \tanh \left(7.25 \times 10^{-7}\text{We}^{1.25}\ \text{Re}_{\text{SL}}^{0.25} \right) \left. \begin{array}{l} 0.1 \leq p_{\text{sys}} \leq 0.4\ \text{MPa} \\ 9.5 \leq D \leq 32\ \text{mm} \\ 370 \leq \text{Re}_{\text{SL}} \leq 6400 \end{array} \right\} \tag{6.4}$$

$$We = \frac{\rho_G U_{SG}^2 D}{\sigma} \left(\frac{\rho_L - \rho_G}{\rho_G} \right)^{0.33} \tag{6.5}$$

Cioncolini and Thome (2012) Correlation They proposed a method to predict liquid entrainment fraction in gas–liquid annular flows. In comparison to Ishii and Mishima (1989), correlation of Cioncolini and Thome (2012) is based on a more comprehensive data set, 2293 data points for circular pipes and 71 data points for noncircular pipes, and is claimed to be valid for $5 \leq D \leq 95$ mm and $0.1 \leq p_{sys} \leq 20$ MPa and for both adiabatic and evaporating flows in vertical upward flow.

Bhagwat and Ghajar (2015b) Correlation They introduced a correction factor $(\xi + 120\cos^2\theta)$ to the Cioncolini and Thome (2012) correlation to improve its accuracy at high system pressures and horizontal and upward inclined pipe orientations. This correction factor embedded into original correlation of Cioncolini and Thome (2012) is expressed by Eq. (6.6) where ξ is calculated from Eq. (6.7). It should be pointed out that the original correlation of Cioncolini and Thome (2012) used a constant value 279.6 in place of the correction factor $(\xi + 120\cos^2\theta)$ in Eq. (6.7). The need of a correction factor for the effect of high system pressure on liquid entrainment fraction is explained in the wok by Bhagwat and Ghajar (2015b). Note that the use of Eq. (6.7) requires system pressure (p_{sys}) with units of MPa. The gas core Weber number and the gas core density are calculated using Eqs. (6.8) and (6.9), respectively:

$$E = \left(1 + (\xi + 120\cos^2\theta)We_c^{-0.8395}\right)^{-2.209} \left.\begin{array}{l} 0.1 \leq p_{sys} \leq 20 \text{ MPa} \\ 5 \leq D \leq 95 \text{ mm} \\ 10 \leq We_c \leq 10^5 \end{array}\right\} \tag{6.6}$$

$$\xi = \left(\begin{array}{ll} 280 & : 0.1 \leq p_{sys} < 10 \text{ MPa} \\ 4637.8 \times p_{sys}^{-1.6} & : p_{sys} \geq 10 \text{ MPa} \end{array} \right. \tag{6.7}$$

$$We_c = \frac{\rho_c U_{SG}^2 D}{\sigma} \tag{6.8}$$

$$\rho_c = \frac{x + E(1 - x)}{(x/\rho_G) + (E(1 - x)/\rho_L)} \tag{6.9}$$

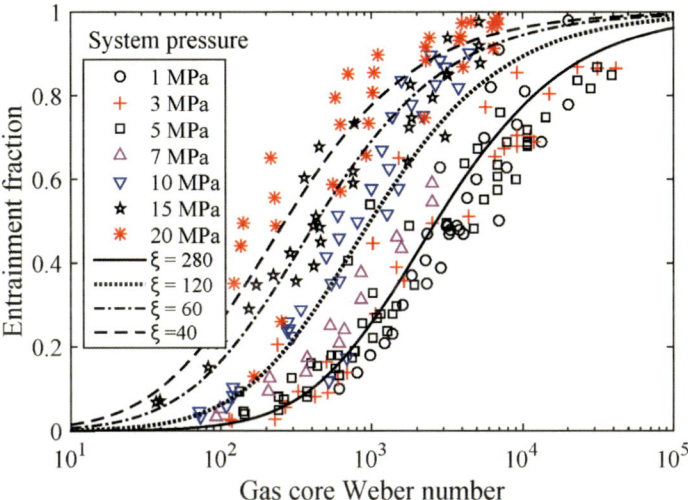

Fig. 6.3 Variation of liquid entrainment fraction as a function of system pressure (trends predicted by Eq. (6.6))

Equation (6.6) needs to be solved using iterative technique since the core density required in calculation of the core Weber number depends upon the liquid entrainment fraction. Cioncolini and Thome (2012) suggested a two-step predictor and corrector method such that the predictor method predicts the liquid entrainment fraction for an initial guess of core density approximately equal to the gas-phase density (i.e., $\rho_c \approx \rho_G$). In the corrector step, the E value from predictor step is used to calculate a new value of gas core density and hence the liquid entrainment fraction. The variation of liquid entrainment fraction as a function of system pressure using the proposed modification by Bhagwat and Ghajar (2015b) to the original Cioncolini and Thome (2012) correlation is illustrated in Fig. 6.3. In this figure, the value of $\xi \approx 280$ is the original fixed value used by Cioncolini and Thome (2012), and the ξ values of 120, 60, and 40 correspond to 10, 15, and 20 MPa of the system pressure, respectively (see Eq. 6.7). Clearly the proposed modification predicts more than 90% of the data points within $\pm 30\%$ deviation for system pressures of 10, 15, and 20 MPa which shifts the trend of E vs. We_c with change in ξ values which is a function of system pressure. From Fig. 6.3 it is evident that at low ($We_c \leq 10$) and high ($We_c \geq 10^5$) gas core Weber numbers, the effect of system pressure on liquid entrainment fraction is small. Figure 6.3 is plotted for vertical upward two-phase flow and can be used as a graphical solution to determine liquid entrainment fraction.

An Illustrative Example: Use of Bhagwat and Ghajar (2015b) Correlation Consider the vertical upward annular two-phase flow of air–water mixture in a 45 mm I.D. smooth pipe. At near atmospheric system pressure, the two-phase mixture flows with a quality (x) and mass flux (G) of 0.25 and 210 kg/m^2, respectively. The physical properties of air and water are as follows: $\rho_G = 1.5$ kg/m^3, $\rho_L = 998$ kg/m^3, and $\sigma = 0.072$ N/m. Determine the liquid entrainment fraction.

Solution
Liquid entrainment fraction (E) can be estimated from Eq. (6.6) using an iterative technique since the core density (ρ_c) required in calculation of the core Weber number (We_c) depends upon the liquid entrainment fraction. Using a two-step predictor and corrector method, first estimate the gas core Weber number (We_c) from Eq. (6.8) using gas-phase density (ρ_G) and superficial gas velocity (U_{SG}) to predict the initial estimate of the liquid entrainment fraction. The superficial gas velocity required in determination of We_c is calculated as $U_{SG} = Gx/\rho_G = 210 \times 0.25/1.5 = 35$ m/s. Then find the gas core density (ρ_c) from Eq. (6.9) using the initial estimate of E, and recalculate core Weber number and liquid entrainment fraction in corrector step. For near atmospheric system pressure $\xi = 280$ from Eq. (6.7).

Predictor Step Initial estimate of gas core Weber number (We_c) using Eq. (6.8).

$$We_c = \frac{\rho_G U_{SG}^2 D}{\sigma} = \frac{1.5 \times 35^2 \times 0.045}{0.072} = 1148.4$$

Initial estimate of the liquid entrainment fraction (E) using Eq. (6.6).

$$E = \left(1 + 280 \times We_c^{-0.8395}\right)^{-2.209} = 0.288$$

Corrector Step Calculate the gas core density (ρ_c) from Eq. (6.9) using the initial estimate of E, and recalculate the gas core Weber number (We_c):

$$\rho_c = \frac{x + E(1 - x)}{(x/\rho_G) + (E(1 - x)/\rho_L)} = \frac{0.25 + 0.288 \times 0.75}{(0.25/1.5) + (0.288 \times 0.75/998)}$$
$$= 2.8 \ (kg/m^3)$$

$$We_c = \frac{\rho_c U_{SG}^2 D}{\sigma} = \frac{2.8 \times 35^2 \times 0.045}{0.072} = 2143.7$$

The final value of liquid entrainment fraction (E) is determined from the recalculated value of the gas core Weber number:

$$E = \left(1 + 280 \times We_c^{-0.8395}\right)^{-2.209} = 0.44$$

Discussion The calculated value of the liquid entrainment fraction (E) can now be used to determine, for example, the liquid film mass flow rate ($\dot{m}_{LF}$) as follows:

$$\dot{m}_{LF} = (1 - E)\dot{m}_L = (1 - 0.44) \times 0.25 = 0.14\,\mathrm{kg/s}$$

where the liquid mass flow rate ($\dot{m}_L$) required in the $\dot{m}_{LF}$ equation is calculated as

$$\dot{m}_L = G(1 - x) \times A = 210(1 - 0.25) \times \pi/4 \times 0.045^2 = 0.25\ \mathrm{kg/s}$$

Chapter 7
Non-Boiling Two-Phase Heat Transfer

Background

Knowledge of non-boiling two-phase heat transfer is important in several applications in oil and gas and chemical engineering industry. In particular, during the production of two-phase hydrocarbon fluids from oil reservoirs and its transportation to the surface processing facilities, the temperature of hydrocarbons drops drastically and is favorable for hydrates formation and wax deposition. Wax deposition can result in problems including reduction of inner tube diameter causing blockage, increase in surface roughness of tube leading to restricted flow line pressure, and decrease in production, which can lead to various mechanical problems. In such situations, the correct knowledge of heat transfer coefficients in two-phase flow is of utmost importance for the purpose of flow assurance in oil and gas industry.

The non-boiling heat transfer coefficient in gas–liquid two-phase flow similar to void fraction and two-phase pressure drop is found to be influenced by several parameters such as pipe orientation and flow rates of individual phases. These influences on the two-phase heat transfer coefficient for different pipe orientations and liquid and gas flow rates will be briefly discussed next. Detailed information on these influences can be obtained from Ghajar and Tang (2010) and Bhagwat and Ghajar (2016, 2017).

Effect of Pipe Inclination and Phase Flow Rates on Heat Transfer Coefficient

As shown in Fig. 7.1, in upward pipe inclinations, the two-phase heat transfer coefficient (h_{TP}) increases linearly with increase in the pipe inclination from horizontal and attains a maximum value at $\theta = +75°$. At low gas and liquid flow rates,

© The Author(s), under exclusive license to Springer Nature Switzerland AG 2020
A. J. Ghajar, *Two-Phase Gas-Liquid Flow in Pipes with Different Orientations*,
SpringerBriefs in Applied Sciences and Technology,
https://doi.org/10.1007/978-3-030-41626-3_7

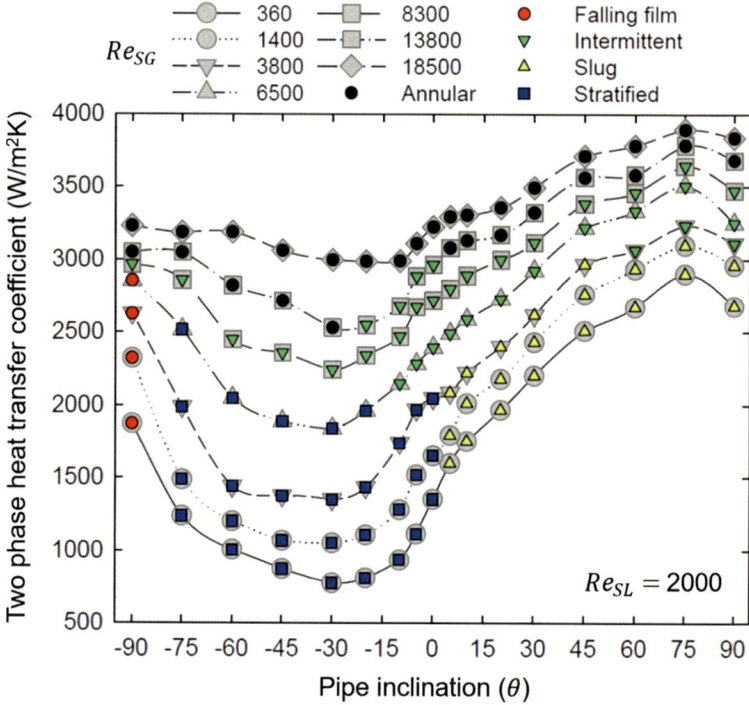

Fig. 7.1 Effect of change in pipe inclination on two-phase heat transfer coefficient at $Re_{SL} = 2000$. (Data measured at the Two-Phase Flow Lab, OSU, Stillwater, OK)

h_{TP} increases by $\approx 55\%$ as pipe is inclined from horizontal toward vertical upward position. For the case of low liquid flow rates and high gas flow rates, this difference between h_{TP} at horizontal and vertical upward flow is reduced to $\approx 6\%$. Note that at moderate liquid flow rates and high gas flow rates, the difference between h_{TP} at horizontal and vertical upward pipe inclination is more substantial. This is because of the combined effect of increase in sweeping action of the disturbance waves at the gas–liquid interface and gradual approach toward uniform cross-sectional distribution of the flow with increase in the liquid flow rate. Both of these conditions favor and aid heat transfer process in two-phase flow.

Similar to the total two-phase pressure drop trends (see Fig. 5.1a), the two-phase heat transfer coefficient at +75° is higher than that at +60° and +90° inclinations. This anomaly is possibly due to the phenomenon of liquid accumulation at pipe bottom wall that obstructs the motion of gas phase and eventually gives rise to the blow through action of the gas phase. This entire process results into better vigorous interaction between the two phases and hence increases the two-phase heat transfer coefficient. Similar observation is reported in literature by Spedding et al. (2000).

In comparison to two-phase total pressure drop (see Fig. 5.1a), in downward pipe inclinations, the two-phase heat transfer coefficient responds sharply to the change in downward pipe inclination. As shown in Fig. 7.1, at $Re_{SL} = 2000$ and the lowest value of $Re_{SG} = 360$, the two-phase heat transfer coefficient initially decreases with decrease in pipe inclination from horizontal and attains a minimum value in the vicinity of $\theta = -30°$ and then increases as the pipe is inclined toward vertical downward position. At $Re_{SL} = 2000$ and varying Re_{SG}, the average decrease in h_{TP} at $\theta \approx -30°$ with reference to the horizontal flow is about 30%. This inflection in the trend of h_{TP} vs. θ appears to be the consequence of slippage at the gas–liquid interface which in turn is a function of pipe inclination. For low gas and liquid flow rates, stratified flow pattern prevails in downward pipe inclinations. The physical structure of stratified flow patterns in downward pipe inclinations is such that there is no coupling between the two phases except the interfacial interaction. At low gas flow rates, the gravity driven nature of the flow increases the residence time of gas phase in the test section and as a result decreases the two-phase heat transfer coefficient. It is interesting to note that the general trends of void fraction show inflection at $\theta \approx -45°$ (see Fig. 4.1); however, the inflection in two-phase heat transfer coefficient variation is at $\theta \approx -30°$. Based on the flow visualization, there appears to be no physical reason behind this discrepancy, and thus more detailed experiments such as measurements of liquid film thickness are deemed necessary for better understanding of this deviation. Similar to upward pipe inclinations, effect of downward pipe inclination on h_{TP} diminishes with increase in gas and liquid flow rates (inertia driven region of two-phase flows). For instance, at $Re_{SL} = 6000$, the highest and average decrease in two-phase heat transfer coefficients are 11% and 3% compared to 47% and 30% at $Re_{SL} = 2000$. As shown in Fig. 7.1, at steeper downward pipe inclinations and low liquid and gas Reynolds numbers, higher heat transfer coefficient is due to better wetting of the pipe periphery due to unstable gas–liquid interface (and hence frequent splashing). This will be explained further in the next section.

Circumferential Variation of Heat Transfer Coefficient with Pipe Inclination

Since the experimental setup described in Chap. 2 measures local values of two-phase heat transfer coefficient, it is also of interest to check the circumferential variation of h_{TP} for fixed values of gas and liquid flow rates/velocities and varying pipe orientations. Figure 7.2 shows circumferential values of two-phase heat transfer coefficient for horizontal and various upward pipe inclinations. Each local h_{TP} is the average of all axial two-phase heat transfer coefficients (at seven axial stations) measured at a fixed circumferential location. It is observed that due to lack of symmetry in upward pipe inclinations, the heat transfer coefficient at pipe bottom is always higher than the top and side walls of the pipe. As the pipe inclination approaches vertical upward position, the difference between the circumferential h_{TP} decreases. In case of vertical upward flow, all four values of h_{TP} (at top, bottom, and two sides of the pipe) are comparable because of the flow symmetry. For low values

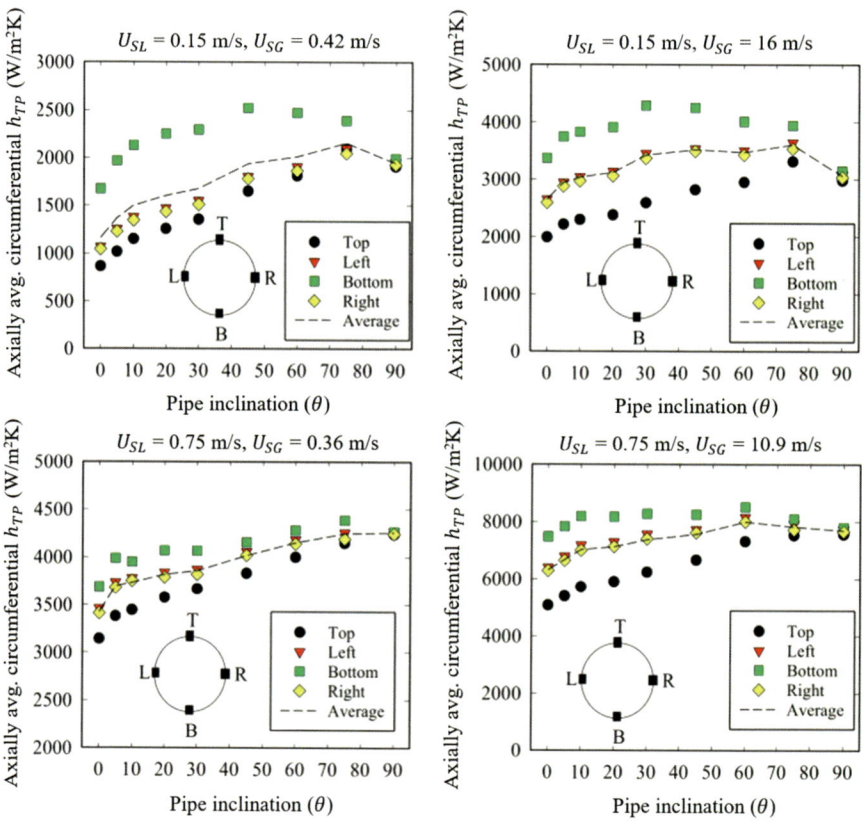

Fig. 7.2 Circumferential variation of two-phase heat transfer coefficient as a function of pipe orientation for fixed values of gas and liquid flow rates. (Data measured at the Two- Phase Flow Lab, OSU, Stillwater, OK)

of U_{SL} and U_{SG}, the average value (axial as well as circumferential) of two-phase heat transfer coefficient is influenced by the heat transfer coefficients at the pipe top and side walls. This situation corresponds to slug/wavy slug flow where bottom wall of the pipe is always in contact with the thick liquid film, while the top and side walls are wetted intermittently due to alternate flow of gas and liquid slugs. With increase in the gas flow rate, for example, at $U_{SG} = 16$ m/s, entire pipe cross section is wetted continuously resulting into the shift of average heat transfer coefficient toward the mean of local h_{TP} at bottom and top wall of the pipe. However, there is still a difference between top and bottom wall heat transfer coefficient (for non-vertical flow) since the liquid film thickness at bottom wall is thicker compared to that at the top wall of the pipe, and hence $h_{TP, \, Bottom} > h_{TP, \, Top}$.

As it was mentioned earlier, at steeper downward pipe inclinations and low liquid and gas Reynolds numbers, higher heat transfer coefficients are observed (see Fig. 7.1). The increase in h_{TP} for steeper downward pipe inclinations could be explained based on the variation in the physical structure of the stratified flow pattern. For these steeper (near vertical and vertical downward) pipe orientations, the flow pattern is stratified and falling film flow, respectively. At $\theta = -60^\circ$, the visual observations show that the gas–liquid interface of stratified flow pattern is unstable such that the liquid phase splashes frequently on the pipe top wall and momentarily bridges the pipe cross section. Barnea et al. (1982) also reported that at steeper pipe inclinations, liquid lumps are torn away from the unstable gas–liquid interface all the way to the top wall of the pipe. They also mentioned that at low to moderate liquid flow rates, the gas–liquid interface in steeper pipe orientations tend to become concave, and the liquid film climbs the tube periphery with increase in the liquid flow rate and pipe orientation. Additionally, at higher gas flow rates, the work of Fukano and Ousaka (1989) reveals that the disturbance waves create a pumping action on the liquid layer causing it to rise in circumferential direction and also carry it in the downstream direction. Thus, the liquid phase moves in both circumferential as well as axial direction and provides mixing and secondary flows. This implies that, compared to near horizontal downward pipe inclinations, a greater fraction of the pipe circumference (wetted perimeter) is in contact with the liquid phase for near vertical downward pipe inclinations and hence allows higher rates of two-phase heat transfer. Maximum value of the two-phase heat transfer coefficient is observed for the case of vertical downward flow. In case of vertical downward flow, entire pipe circumference is wetted by a thin axisymmetric liquid film (falling film flow), whereas the stratified flow in near vertical downward pipe inclinations experiences asymmetric and thick liquid film that partially wets the pipe circumference. Thus, the presence of thin liquid film offers lesser thermal resistance and hence results into highest heat transfer coefficient. The three different forms of stratified flow with variation in the liquid film thickness and the corresponding circumferential variation of two-phase heat transfer coefficient for different downward pipe inclinations are depicted in Fig. 7.3. Each local h_{TP} is the average of all axial two-phase heat transfer coefficients (at seven axial stations) measured at a fixed circumferential location. The variation in circumferential heat transfer coefficients is in agreement with the aforementioned circumferential variation of liquid film/layer thickness as a function of pipe inclination. For $\theta = -5^\circ$, the local two-phase heat transfer coefficient is highest at pipe bottom wetted by a liquid film while that the pipe top wall is smallest due to the absence of liquid film. As the pipe inclination approaches vertical downward position, all the local two-phase heat transfer coefficients are comparable indicating the presence of axisymmetric two-phase flow.

Modeling of Non-Boiling Two-Phase Heat Transfer

In two-phase flow literature, majority of the documented non-boiling two-phase heat transfer modeling methods are limited to specific pipe orientations, fluid combinations, and most importantly flow patterns. The subsea as well as surface pipelines carrying mixture of oil and gas together are usually laid in undulated form having near horizontal inclinations, and the existence of a specific flow pattern is uncertain.

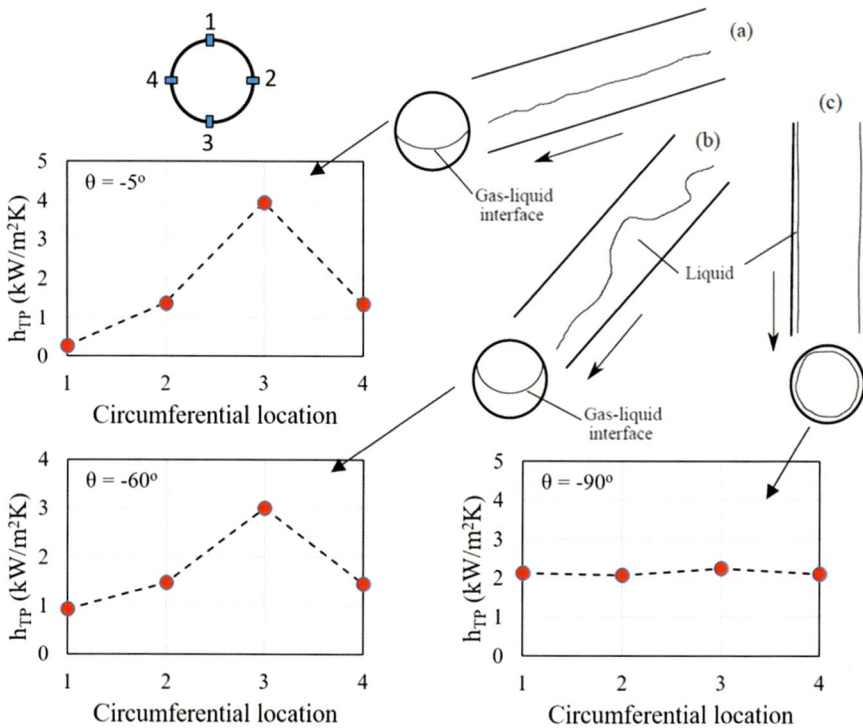

Fig. 7.3 Variation in liquid film thickness and corresponding local two-phase heat transfer coefficients for (**a**) near horizontal, (**b**) steep, and (**c**) vertical downward pipe inclinations ($\mathrm{Re}_{\mathrm{SL}} = 2000$, $\mathrm{Re}_{\mathrm{SG}} = 280$). (Data measured at the Two-Phase Flow Lab, OSU, Stillwater, OK)

For such a scenario, the use of flow pattern and pipe orientation-specific two-phase heat transfer correlations may predict significant deviations from the actual conditions. To address this issue, Ghajar and coworkers in a series of papers have extensively studied the phenomenon of non-boiling two-phase heat transfer in pipes at different orientations; see, for example, Kim et al. (1999, 2000) and Kim and Ghajar (2006). Ghajar and Tang (2009) proposed a robust correlation for two-phase heat transfer coefficient (h_{TP}) as presented by Eq. (7.1). Their correlation is based on 986 experimental data points for different pipe orientations, fluid combinations, and flow patterns:

$$h_{\mathrm{TP}} = h_{\mathrm{L}} F_{\mathrm{p}} \left[1 + 0.55 \left(\frac{x}{1-x} \right)^{0.1} \left(\frac{1 - F_{\mathrm{p}}}{F_{\mathrm{p}}} \right)^{0.4} \left(\frac{\mathrm{Pr}_{\mathrm{G}}}{\mathrm{Pr}_{\mathrm{L}}} \right)^{0.25} \left(\frac{\mu_{\mathrm{L}}}{\mu_{\mathrm{G}}} \right)^{0.25} I^{0.25} \right] \quad (7.1)$$

$$F_p = (1 - \alpha_G) + \alpha_G F_S^2 \tag{7.2}$$

$$F_S = \frac{2}{\pi} \arctan\left(\sqrt{\frac{\rho_G(U_G - U_L)^2}{gD(\rho_L - \rho_G)}}\right) \tag{7.3}$$

$$I = 1 + Eo|\sin\theta| \tag{7.4}$$

The flow pattern factor (F_p) given by Eq. (7.2) makes the Ghajar and Tang (2009) correlation independent of the flow patterns and depends on shape factor F_S and void fraction (α_G). The shape factor (F_S) given by Eq. (7.3) considers the effect of the shape of gas–liquid interface on two-phase heat transfer. The shape factor (F_S) is applicable for slip ratios equal or greater than one ($S \geq 1$), which is common in gas–liquid flow, and represents the shape changes of the gas–liquid interface by the force acting on the interface due to the relative momentum and gravitational forces. The void fraction (α_G) in Eq. (7.2) should be calculated using the simple correlations of Woldesemayat and Ghajar (2007), using Eqs. (4.11) and (4.12). The expression for the inclination factor (I), given by Eq. (7.4), includes representation of the relative forces acting on the liquid phase in the flow direction due to the momentum and buoyancy forces. The inclination factor accounts for the effect of pipe orientation on h_{TP} and is defined in terms of Eötvös number:

$$Eo = \frac{(\rho_L - \rho_G)gD^2}{\sigma} \tag{7.5}$$

The single-phase heat transfer coefficient (h_L) required in Eq. (7.1) is based on in situ Reynolds number. All thermophysical properties required are calculated at bulk mean temperature except for the μ_w that is calculated at the pipe wall temperature. Note that the in situ liquid-phase Reynolds number used in Eq. (7.6) for h_L is defined in terms of the square root of liquid holdup as shown in Eq. (7.7):

$$h_L = 0.027 \, Re_L^{0.8} Pr_L^{0.33} \left(\frac{k_L}{D}\right)\left(\frac{\mu_L}{\mu_w}\right)^{0.14} \tag{7.6}$$

$$Re_L = \frac{4\dot{m}_L}{\pi D \mu_L \sqrt{1 - \alpha_G}} = \frac{G(1-x)D}{\mu_L \sqrt{1 - \alpha_G}} \tag{7.7}$$

Comparison of Ghajar and Tang (2009) Correlation with Experimental Data The general two-phase heat transfer correlation of Ghajar and Tang (2009), Eq. (7.1), is verified against 986 experimental data points (176 for horizontal flow, 555 for inclined flow, and 255 for upward vertical flow) from five

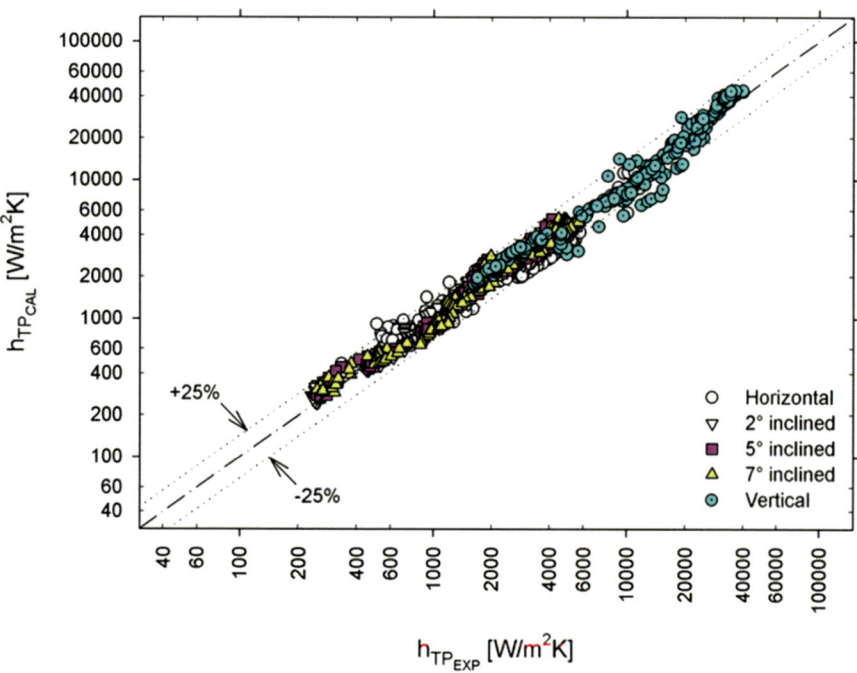

Fig. 7.4 Comparison of the predictions by Eq. (7.1) with all 986 experimental data points for different inclination angles, gas–liquid combinations, and flow patterns

different sources with different experimental facilities. The data covered different pipe inclination angles, gas–liquid combinations (air–water, air–silicone, helium–water, Freon-12–water), and all flow patterns with the exception of stratified flow. Their correlation is applicable over the following range of parameters: $750 \leq Re_{SL} \leq 1.27 \times 10^5$, $14 \leq Re_{SG} \leq 21 \times 10^5$, $0.01 \leq Pr_G/Pr_L \leq 0.15$, $0.0036 \leq \mu_G/\mu_L \leq 0.026$, 910 kg/m$^3 \leq \rho_L \leq 1210$ kg/m^3, and $0° \leq \theta \leq +90°$. Figure 7.4 shows the comparison of their correlation with the entire set of experimental data. Their correlation predicted 90% of the experimental data within $\pm25\%$, confirming the robustness of Eq. (7.1) for prediction of two-phase heat transfer coefficients.

Application of Reynolds Analogy to Non-Boiling Two-Phase Heat Transfer

The concept of Reynolds analogy that relates the friction coefficient or the frictional pressure drop with heat transfer in single-phase flow can also be extended to the case of two-phase flow. Using the concept of Reynolds analogy, Tang and Ghajar (2011) have developed a correlation to predict non-boiling two-phase heat transfer coefficient from the knowledge of two-phase frictional pressure drop in horizontal and upward inclined flows. Their correlation correlates the two-phase to single-phase heat transfer coefficient ratio to the two-phase frictional multiplier (ratio of two-phase to single-phase pressure drop). The physical form of their correlation is given by Eq. (7.8). Tang (2011) suggested that the exponents of F_p and Φ_L could vary between 0.1 and 0.5 depending upon the pipe orientation and fluid properties; however, the general structure of Reynolds analogy-based correlation remains unaltered. In Eq. (7.8), the exponent (0.3) of F_p and Φ_L is based on the overall performance of Tang and Ghajar (2011) model for various pipe inclinations and fluid combinations. The same experimental data set used for the development of Eq. (7.1) was used for the development of Eq. (7.8). The Reynolds analogy-based two-phase heat transfer correlation of Tang and Ghajar (2011), Eq. (7.8), predicted 85% of the experimental data within ±30%:

$$h_{TP} = h_L F_p^{0.3} \left(\frac{\dot{m}_L}{\dot{m}}\right) \left(\frac{\rho_L}{\rho_M}\right)^{0.5} \Phi_L^{0.3} \tag{7.8}$$

$$\rho_M = \rho_L(1 - \alpha_G) + \rho_G \alpha_G \tag{7.9}$$

$$\left(\frac{dp}{dz}\right)_f = \Phi_L^2 \left(\frac{dp}{dz}\right)_L \quad \text{where} \quad \left(\frac{dp}{dz}\right)_L = \frac{2f_L G^2 (1 - x)^2}{D\rho_L} \tag{7.10}$$

In Eq. (7.8), the two-phase mixture density (ρ_M) is calculated from Eq. (7.9) using void fraction. The void fraction required in calculation of ρ_M is calculated using Woldesemayat and Ghajar (2007) correlation (Eqs. 4.11 and 4.12). F_p is calculated from Eq. (7.2), and h_L is calculated from Eq. (7.6) using the Reynolds number (Re_{SL}) based on superficial liquid velocity (see Table 1.1). Note that Φ_L used in Eq. (7.8) is the square root of the two-phase frictional multiplier calculated using Eq. (7.10). In Eq. (7.10), $(dp/dz)_L$ and $(dp/dz)_f$ are the single-phase liquid and two-phase frictional pressure drop, respectively. The friction factor (f_L) required for the single-phase liquid pressure drop calculation can be calculated using any appropriate friction factor correlation such as correlation of Blasius (1913) or Churchill (1977). In case the two-phase frictional pressure drop $(dp/dz)_f$ is not known/measured, then suitable correlation valid for given two-phase flow conditions such as the correlations recommended by Ghajar and Bhagwat (2017) may be used to determine Φ_L.

For vertical downward flow, Bhagwat et al. (2012) pointed out that the two-phase heat transfer coefficient is less than that measured in vertical upward flow at similar mass flow rates, while the two-phase pressure drop is considerably higher in the former orientation than the latter. Thus, Tang and Ghajar (2011) Reynolds analogy-based heat transfer correlation, Eq. (7.8), with fixed exponents does not give satisfactory performance in prediction of two-phase heat transfer coefficient for vertical downward flow. Bhagwat et al. (2012) proposed the following form of the

Reynolds analogy-based correlation that can be used to predict non-boiling heat transfer coefficient for vertical downward flow:

$$h_{TP} = h_L \Phi_L^{0.55} \tag{7.11}$$

In Eq. (7.11), the single-phase heat transfer coefficient (h_L) and the two-phase frictional multiplier (Φ_L) are calculated the same way as that of Eq. (7.8). Equation (7.11) is based on the 165 data points measured at the OSU two-phase flow lab in a vertically downward-oriented 0.0127 m stainless steel pipe using air–water fluid combination. This correlation predicts more than 90% of the measured data points within ±20% error bands and is applicable for $1400 \leq Re_{SL} \leq 29{,}000$, and $90 \leq Re_{SG} \leq 18{,}000$.

It should be noted the physical structure of both Ghajar and Tang (2009) general heat transfer correlation, Eq. (7.1), and Tang and Ghajar (2011) Reynolds analogy-based heat transfer correlation, Eq. (7.8), is modular and robust, and hence the application of these correlations can be potentially extended to a wider range of two-phase flow conditions by modifying the empirical multiplying factors and exponents. Both of these correlations, Eqs. (7.1) and (7.8), require accurate information of void fraction. Ghajar and Bhagwat (2013, 2014b) have evaluated several void fraction correlations available in the literature against air–water data (5162 data points for horizontal, upward, and downward inclinations) and refrigerant-vapor (645 data points for horizontal flow). They found the performance of Woldesemayat and Ghajar (2007) correlation (see Eqs. (4.11) and (4.12)) or a more recent correlation by Bhagwat and Ghajar (2014) for a wide range of gas–liquid two-phase flow (see Eqs. (4.13, 4.14, 4.15, 4.16, 4.17, 4.18, and 4.19) are acceptable for general application with requisite accuracy. For more specific flow conditions, other top-performing void fraction correlations shortlisted by Ghajar and Bhagwat (2013, 2014b) may be used.

An Illustrative Example: Use of the General and Reynolds Analogy-Based Heat Transfer Correlations Consider the vertical upward two-phase flow of air and silicone oil in a 12 mm I.D. stainless steel pipe having a surface roughness of 0.002 mm. The mass flow rate of gas and liquid phase is 0.0015 kg/s and 0.9 kg/s, respectively. The fluid thermophysical properties may be taken as follows: $\rho_G = 1.2$ kg/m^3, $\rho_L = 920$ kg/m^3, $\mu_G = 18.4 \times 10^{-6}$ Pa·s, $\mu_L = 0.005$ Pa·s, $\mu_w = 0.004$ Pa·s, $\sigma = 0.02$ N/m, $k_L = 0.12$ W/mK, $Pr_G = 0.71$, and $Pr_L = 64$. For a measured void fraction of $\alpha_G = 0.5$, determine the two-phase heat transfer coefficient for the flow of air and silicon oil using (1) the general heat transfer correlation of Ghajar and Tang (2009), Eq. (7.1), and (2) Reynolds analogy-based heat transfer correlation of Tang and Ghajar (2011), given by Eq. (7.8).

Solution
1. Calculation of h_{TP} using Ghajar and Tang (2009) general heat transfer correlation:
   Referring to Table 1.1, the superficial gas (U_{SG}) and liquid (U_{SL}) velocities and the two-phase flow quality (x) are first calculated as follows:

$$U_{SG} = \frac{4\dot{m}_G}{\pi D^2 \rho_G} = \frac{4 \times 0.0015}{\pi \times 0.012^2 \times 1.2} = 11.05 \ \text{m/s}$$

$$U_{SL} = \frac{4\dot{m}_L}{\pi D^2 \rho_L} = \frac{4 \times 0.9}{\pi \times 0.012^2 \times 920} = 8.65 \ \text{m/s}$$

$$x = \frac{\dot{m}_G}{\dot{m}_G + \dot{m}_L} = \frac{0.0015}{0.0015 + 0.9} = 0.00166$$

From the given information of void fraction, the actual velocity of each phase is calculated as follows. In case the void fraction is not known, Woldesemayat and Ghajar (2007) explicit void fraction correlation, Eqs. (4.11) and (4.12), or Bhagwat and Ghajar (2014) implicit void fraction correlation, Eqs. (4.13, 4.14, 4.15, 4.16, 4.17, 4.18, and 4.19), can be used for calculation of the void fraction. Refer to the example problem in Chap. 4 for the details of the void fraction calculations using either one of the correlations:

$$U_G = \frac{U_{SG}}{\alpha_G} = \frac{11.05}{0.5} = 22.1 \ \text{m/s}$$

$$U_L = \frac{U_{SL}}{1 - \alpha_G} = \frac{8.65}{1 - 0.5} = 17.3 \ \text{m/s}$$

Now, the shape factor is calculated from Eq. (7.3):

$$F_S = \left[\frac{2}{\pi} \arctan \left(\sqrt{\frac{\rho_G (U_G - U_L)^2}{g D (\rho_L - \rho_G)}} \right) \right]$$

$$= \left[\frac{2}{\pi} \arctan \left(\sqrt{\frac{1.2 \times (22.1 - 17.3)^2}{9.81 \times 0.012 \times (920 - 1.2)}} \right) \right] = 0.298$$

The flow pattern factor is obtained from Eq. (7.2):

$$F_p = (1 - \alpha_G) + \alpha_G F_S^2 = (1 - 0.5) + 0.5 \times 0.298^2 = 0.544$$

The inclination factor for vertical flow is calculated by Eq. (7.4):

$$I = 1 + \frac{(\rho_L - \rho_G)gD^2}{\sigma} |\sin\theta| = 1 + \frac{(920 - 1.2) \times 9.81 \times 0.012^2}{0.02} = 65.9$$

Next, the in situ Reynolds number based on void fraction is calculated from Eq. (7.7):

$$Re_L = \frac{4\dot{m}_L}{\pi D \mu_L \sqrt{1 - \alpha_G}} = \frac{4 \times 0.9}{\pi \times 0.012 \times 0.005 \times \sqrt{1 - 0.5}} = 27009.5$$

The single-phase heat transfer coefficient is then calculated using Eq. (7.6):

$$h_L = 0.027 \, Re_L^{0.8} Pr_L^{0.33} \left(\frac{k_L}{D}\right) \left(\frac{\mu_L}{\mu_w}\right)^{0.14}$$

$$= 0.027 \times 27009.5^{0.8} \times 64^{0.33} \times \left(\frac{0.12}{0.012}\right) \left(\frac{0.005}{0.004}\right)^{0.14} = 3856 \ \text{W/m}^2\text{-K}$$

Using the general two-phase heat transfer correlation of Ghajar and Tang (2009) given by Eq. (7.1), h_{TP} is calculated as

$$h_{TP} = h_L F_p \left[1 + 0.55 \left(\frac{x}{1-x}\right)^{0.1} \left(\frac{1-F_p}{F_p}\right)^{0.4} \left(\frac{Pr_G}{Pr_L}\right)^{0.25} \left(\frac{\mu_L}{\mu_G}\right)^{0.25} I^{0.25}\right]$$

$$= 3856 \times 0.544 \left[1 + 0.55 \left(\frac{0.00166}{1-0.00166}\right)^{0.1} \left(\frac{1-0.544}{0.544}\right)^{0.4} \left(\frac{0.71}{64}\right)^{0.25} \times \left(\frac{0.005}{18.4 \times 10^{-6}}\right)^{0.25} 65.9^{0.25}\right]$$

$$= 4224 \ \text{W/m}^2\text{-K}$$

2. Calculation of h_{TP} using Tang and Ghajar (2011) Reynolds analogy-based heat transfer correlation: To use the Reynolds analogy-based heat transfer correlation, we first need to find the two-phase frictional pressure drop and hence the two-phase frictional multiplier $\left(\Phi_L^2\right)$. The following steps calculate the two-phase pressure drop using method of Lockhart and Martinelli (1949).

The Reynolds number of each phase and associated friction factors are calculated as shown below. Since the two-phase flow is through steel pipe having a roughness of 0.002 mm, we need to use appropriate correlation such as Colebrook (1939) or Churchill (1977) to account for the effect of pipe wall surface roughness on friction factor. Here the Colebrook (1939) fanning friction factor correlation was used:

$$Re_{SG} = \frac{\rho_G U_{SG} D}{\mu_G} = \frac{1.2 \times 11.05 \times 0.012}{18.4 \times 10^{-6}} = 8,648 \Rightarrow f_G = 0.0081$$

$$\text{Re}_{SL} = \frac{\rho_L U_{SL} D}{\mu_L} = \frac{920 \times 8.65 \times 0.012}{0.005} = 19,099 \Rightarrow f_L = 0.0063$$

The Colebrook (1939) fanning friction factor correlation with $\varepsilon/D = 1.67 \times 10^{-4}$ was used for calculation of gas-phase friction factor (f_G), and liquid-phase friction factor (f_L) is as shown here:

$$\frac{1}{\sqrt{f}} = -4.0 \ \log \left(\frac{\varepsilon/D}{3.7} + \frac{1.256}{\text{Re} \ \sqrt{f}} \right)$$

Now the single-phase pressure drop due to flow of gas and liquid phase is calculated by

$$\left(\frac{dp}{dz} \right)_G = \frac{2 f_G \rho_G U_{SG}^2}{D} = \frac{2 \times 0.0081 \times 1.2 \times 11.05^2}{0.012} = 197.8 \ \text{Pa/m}$$

$$\left(\frac{dp}{dz} \right)_L = \frac{2 f_L \rho_L U_{SL}^2}{D} = \frac{2 \times 0.0063 \times 920 \times 8.65^2}{0.012} = 72,278.5 \ \text{Pa/m}$$

The Lockhart and Martinelli (1949) parameter X given by Eq. (5.21) is calculated using single-phase pressure drop as shown here:

$$X = \sqrt{\frac{(dp/dz)_L}{(dp/dz)_G}} = \sqrt{\frac{72,278.5}{197.8}} = 19.11$$

Since single-phase flow of both phases is in turbulent region, use $C = 20$ from Table 5.1, and the two-phase frictional multiplier using method of Lockhart and Martinelli (1949) is calculated from Eq. (5.20) as shown here:

$$\Phi_L^2 = 1 + \frac{C}{X} + \frac{1}{X^2} = 1 + \frac{20}{19.11} + \frac{1}{19.11^2} = 2.05 \Rightarrow \Phi_L = 1.43$$

Two-phase mixture density based on void fraction is calculated using Eq. (7.9) as follows:

$$\rho_M = \alpha_G \rho_G + (1 - \alpha_G)\rho_L = 0.5 \times 1.2 + (1 - 0.5) \times 920 = 460.6 \ \text{kg/m}^3$$

Now using Eq. (7.8), two-phase heat transfer coefficient can be found. The value of h_L based on Re_{SL} is calculated from Eq. (7.6). The value of $F_p = 0.544$ was calculated in Part (1) of this problem:

$$h_L = 0.027 \, \mathrm{Re}_{SL}^{0.8} \mathrm{Pr}_L^{0.33} \left(\frac{k_L}{D}\right) \left(\frac{\mu_L}{\mu_w}\right)^{0.14}$$

$$= 0.027 \times 19099^{0.8} \times 64^{0.33} \times \left(\frac{0.12}{0.012}\right)\left(\frac{0.005}{0.004}\right)^{0.14} = 2922.6 \ \mathrm{W/m^2\text{-}K}$$

Finally, the two-phase heat transfer coefficient using Reynolds analogy-based correlation is found using Eq. (7.8):

$$h_{TP} = h_L F_p^{0.3} \left(\frac{\dot{m}_L}{\dot{m}}\right)\left(\frac{\rho_L}{\rho_M}\right)^{0.5} \Phi_L^{0.3}$$

$$= 2922.6 \times 0.544^{0.3} \left(\frac{0.9}{0.9015}\right)\left(\frac{920}{460.6}\right)^{0.5} 1.43^{0.3}$$

$$= 3824 \ \mathrm{W/m^2\text{-}K}$$

Discussion Rezkallah and Sims (1987) made two-phase heat transfer measurements for similar experimental conditions and found that $h_{TP} \approx 3900$ W/m^2-K. The general heat transfer correlation of Ghajar and Tang (2009), Eq. (7.1), overpredicted the measured value of h_{TP} by about 8%, and the Reynolds analogy-based heat transfer correlation of Tang and Ghajar (2011), Eq. (7.8), underpredicted the measured value of h_{TP} by about 2%. The general heat transfer correlation and Reynolds analogy-based heat transfer correlation are sensitive to the void fraction (through F_P and ρ_M). Note that in case the two-phase frictional pressure drop is not known, the accuracy of the Reynolds analogy concept depends on the choice of two-phase frictional multiplier correlation appropriate to the two-phase flow under consideration.

References

Abdulkadir M, Hernandez-Perez V, Sharaf S, Lowndes IS, Azzopardi BJ (2010) Experimental investigation of phase distributions of two-phase air–silicone oil flow in a vertical pipe. World Acad Sci Eng Technol 37:52–59

Abduvayat P, Manabe R, Arihara N (2003) Effects of pressure and pipe diameter on gas-liquid two-phase flow behavior in pipelines, SPE Annual Technical Conference, SPE 84229

Barnea D (1986) Transition from annular and from dispersed bubble flow – unified models for the whole range of pipe inclinations. Int J Multiphase Flow 12:733–744

Barnea D (1987) A unified model for predicting flow pattern transitions for the whole range of pipe inclinations. Int J Multiphase Flow 13:1–12

Barnea D, Shoham O, Taitel Y (1982) Flow pattern transition for downward inclined two-phase flow: horizontal to vertical. Chem Eng Sci 37:735–740

Beggs HD (1972) An experimental study of two-phase flow in inclined pipes, Ph.D. Thesis, University of Tulsa

Beggs HD, Brill JP (1973) A study of two-phase flow in inclined pipes. J Pet Technol 25 (5):607–617

Bendiksen KH (1984) An experimental investigation of the motion of long bubbles in inclined tubes. Int J Multiphase Flow 10:467–483

Bhagwat SM (2015) Experimental measurements and modeling of void fraction and pressure drop in upward and downward inclined non-boiling gas-liquid two-phase flow, Ph.D. Thesis, Oklahoma State University

Bhagwat SM, Ghajar AJ (2014) A flow pattern independent drift flux model based void fraction correlation for a wide range of gas-liquid two-phase flow. Int J Multiphase Flow 59:186–205

Bhagwat SM, Ghajar AJ (2015a) An empirical model to predict the transition between stratified and non-stratified gas-liquid two-phase flow in horizontal and downward inclined pipes. Heat Transfer Eng. 36(18):1489–1498

Bhagwat SM, Ghajar AJ (2015b) Modified liquid entrainment fraction correlation for varying pipe orientation and system pressure. Int J Multiphase Flow 74:1–4

Bhagwat SM, Ghajar AJ (2016) Experimental investigation of non-boiling gas-liquid two-phase flow in upward inclined pipes. Exp Thermal Fluid Sci 79:301–318

Bhagwat SM, Ghajar AJ (2017) Experimental investigation of non-boiling gas-liquid two-phase flow in downward inclined pipes. Exp Thermal Fluid Sci 89:219–237

Bhagwat SM, Mollamahmutoglu M, Ghajar AJ (2012) Experimental investigation and empirical analysis of non-boiling gas-liquid two-phase heat transfer in vertical downward pipe orientation, Proceedings of ASME 2012 Summer Heat Transfer Conference, 2:349–359

© The Author(s), under exclusive license to Springer Nature Switzerland AG 2020
A. J. Ghajar, *Two-Phase Gas-Liquid Flow in Pipes with Different Orientations*,
SpringerBriefs in Applied Sciences and Technology,
https://doi.org/10.1007/978-3-030-41626-3

Blasius H (1913) Das Anhlichkeitsgesetz bei Reibungsvorgangen in Flussikeiten, Gebiete Ingenieurw, 134

Bowers CD, Hrnjak PS (2010) Determination of void fraction in separated two-phase flows using optical techniques, International Refrigeration and Air-Conditioning Conference, Purdue University, pp 2293–2302

Chen XT, Cai XD, Brill JP (1997) Gas liquid stratified wavy flow in horizontal pipelines. J Energy Resour Technol 119:209–216

Cheng SC, Wong YL, Groeneveld DC (1988) CHF prediction for horizontal flow. In: International symposium on phase change heat transfer, Chonqing, pp 211–215

Chisholm D (1967) A therotical basis for the Lockhart-Martinelli correlation for two-phase flow. Int J Heat Mass Transf 10:1767–1778

Chisholm D (1973) Pressure gradients due to the friction during the flow of evaporating two-phase mixtures in smooth tubes and channels. Int J Heat Mass Transf 16:347–358

Churchill SW (1977) Friction factor equation spans all fluid-flow regimes. Chem Eng J 7:91–92

Cioncolini A, Thome JR (2012) Entrained liquid fraction prediction in adiabatic and evaporating annular two-phase flow. Nucl Eng Des 243:200–213

Colebrook CF (1939) Turbulent flow in pipes, with particular reference to the transition between the smooth and rough pipe laws. J Inst Civil Eng 11:1938–1939

Cook W (2008) An experimental apparatus for measurement of pressure drop and void fraction and non-boiling two-phase heat transfer and flow visualization in pipes for all inclinations, M.S. Thesis, Oklahoma State University

Crawford TJ, Weinberger CB, Weisman J (1985) Two-phase flow patterns and void fractions in downward flow part I: steady state flow patterns. Int J Multiphase Flow 11:761–782

Das G, Das PK, Purohit NK, Mitra AK (2002) Liquid holdup in concentric annuli during cocurrent gas–liquid upflow. Can J Chem Eng 80:153–157

Dix GE (1971) Vapor void fractions for forced convection with subcooled boiling at low flow rates, Report NEDO-10491, General Electric Co

Ducoulombier M, Colasson S, Bonjour J, Haberschill P (2011) Carbon dioxide flow boiling in a single microchannel – part I: pressure drops. Exp Thermal Fluid Sci 35:581–596

Fang XD, Xu Y, Zhou ZR (2011) New correlations of single-phase friction factor for turbulent pipe flow and evaluation of existing single-phase friction factor correlations. Nucl Eng Des 241:897–902

Fukano T, Furukawa T (1998) Prediction of the effects of liquid viscosity on interfacial shear stress and frictional pressure drop in vertical upward gas-liquid annular flow. Int J Multiphase Flow 24:587–603

Fukano T, Ousaka A (1989) Prediction of the circumferential distribution of film thickness in horizontal and near-horizontal gas–liquid annular flows. Int J Multiphase Flow 15:403–419

Ghajar AJ, Bhagwat SM (2013) Effect of void fraction and two-phase dynamic viscosity models on prediction of hydrostatic and frictional pressure drop in vertical upward gas-liquid two-phase flow. Heat Transfer Eng 34(13):1044–1059

Ghajar AJ, Bhagwat SM (2014a) Non-boiling gas-liquid two-phase flow phenomenon in near horizontal, upward and downward inclined pipe orientations. Int J Mech Aerosp Ind Mechatron Eng 8:1039–1053

Ghajar AJ, Bhagwat SM (2014b) Flow patterns, void fraction and pressure drop in gas-liquid two-phase flow at different pipe orientations. In: Frontiers and progress in multiphase flow. Springer Int. Publishing, Cham, Chapter 4, pp 157–212

Ghajar AJ, Bhagwat SM (2017) Gas-liquid flow in ducts. In: Michaelides EE, Crowe CT, Schwarzkopf JD (eds) Handbook of multiphase flow, 2nd edn. CRC Press/Taylor & Francis, Boca Rotan, pp 287–356, Chapter 3

Ghajar AJ, Kim J (2006) Calculation of local inside wall convective heat transfer parameters from measurements of the local outside wall temperatures along an electrically heated circular tube. In: Kutz M (ed) Heat transfer calculations. McGraw Hill, New York, pp 3–27

Ghajar AJ, Tang CC (2009) Advances in void fraction, flow pattern maps and non-boiling heat transfer two-phase flow in pipes with various inclinations. Adv Multiphase Flow Heat Transfer 1:1–52

Ghajar AJ, Tang CC (2010) Importance of non-boiling two phase flow heat transfer in pipes for industrial applications. Heat Transfer Eng 31(9):711–732

Godbole PV, Tang CC, Ghajar AJ (2011) Comparison of void fraction correlations for different flow patterns in upward vertical two-phase flow. Heat Transfer Eng 32(10):843–860

Gokcal B (2008) An experimental and theoretical investigation of slug flow for high oil viscosity in horizontal pipes, Ph.D. Thesis. University of Tulsa

Gokcal B, Al-Sarkhi A, Sarica C (2009) Effects of high oil viscosity on drift velocity for horizontal and upward inclined pipes. SPE Proj Fac & Const 4:32–40

Gomez L, Shoham O, Schmidt Z, Choshki R, Northug T (2000) Unified mechanistic model for steady state two-phase flow: horizontal to upward vertical flow. SPE 5:339–350

Hamersma PJ, Hart J (1987) A pressure drop correlation for gas-liquid pipe flow with a small liquid holdup. Chem Eng Sci 42:1187–1196

Haramathy TZ (1960) Velocity of large drops and bubbles in media of infinite or restricted extent. AICHE J 6:281–288

Hart J, Hamersma PJ, Fortuin JMH (1989) Correlations predicting frictional pressure drop and liquid holdup during horizontal gas-liquid pipe flow with a small liquid holdup. Int J Multiphase Flow 15:947–964

Hashemi A, Kim JH, Sursock JP (1986) Effect of diameter and geometry on two-phase flow regimes and carryover in model PWR hot leg, Eighth international heat transfer conference, pp 2443–2451

Hewitt GF, Lacey PMC (1965) The breakdown of the liquid film in annular two-phase flow. Int J Heat Mass Transf 8:781–791

Hewitt GF, Roberts DN (1969) Studies of two-phase flow patterns by simultaneous x-ray and flash photography, Technical Report AERE-M 2159, Atomic Energy Research Establishment

Hewitt G, Lacey PMC, Nicholls B (1965) Transitions in film flow in vertical tube, Proceedings of two-phase flow symposium, Exeter, U. K., Paper B4 2

Hewitt GF, Martin CJ, Wilkes NS (1985) Experimental and modeling studies of churn-annular flow in the region between flow reversal and the pressure drop minimum. Physicochemical Hydrodynamics 6:69–86

Hibiki T, Ishii M (2003) One-dimensional drift flux model and constitutive equations for relative motion between phases in various two-phase flow regimes. Int J Heat Mass Transf 46:4935–4948

Hills JH (1976) The operation of a bubble column at high throughputs part 1: gas holdup measurements. Chem Eng J 12:89–99

Hlaing ND, Sirivat A, Siemanond K, Wilkes JO (2007) Vertical two-phase flow regimes and pressure gradients: effect of viscosity. Exp Thermal Fluid Sci 31:567–577

Inoue Y (2001) Measurement of interfacial area concentration of gas–liquid two-phase flow in a large diameter pipe, M.S. Thesis, Graduate School of Energy Science, Kyoto University

Inoue A, Kurosu T, Aoki T, Yagi M, Misutake T, Morooka S (1995) Void fraction distribution in BWR fuel assembly and evaluation of subchannel code. J Nucl Sci Technol 32:629–640

Ishii M (1977) One-dimensional drift flux model and constitutive equations for relative motion between phases in various two-phase flow regimes, Argonne National Laboratory, pp 77–47

Ishii M, Grolmes MA (1975) Inception criteria for droplet entrainment in two-phase concurrent film flow. AICHE J 21:308–318

Ishii M, Mishima K (1989) Droplet entrainment correlation in annular two-phase flow. Int J Heat Mass Transf 32:1835–1846

Jeyachandra BC (2011) Effect of pipe inclination on flow characteristics of high viscosity oil gas two-phase flow, Ph.D. Thesis, University of Tulsa

Kaji M, Azzopardi BJ (2010) The effect of pipe diameter on the structure of gas-liquid flow in vertical pipes. Int J Multiphase Flow 36:303–313

Kandlikar SG (2002) Fundamental issues related to flow boiling in minichannels and microchannels. Exp Thermal Fluid Sci 26:389–407

Kataoka I, Ishii M (1987) Drift flux model for large diameter pipe and new correlation for pool void fraction. Int J Heat Mass Transf 30:1927–1939

Keinath B (2012) Void fraction, pressure drop and heat transfer in high pressure condensing flows through microchannels, Ph.D. Thesis, Georgia Institute of Technology

Kim J, Ghajar AJ (2006) A general heat transfer correlation for non-boiling gas-liquid flow with different flow patterns in horizontal pipes. Int J Multiphase Flow 32:447–465

Kim D, Ghajar AJ, Dougherty RL, Ryali VK (1999) Comparison of 20 two-phase heat transfer correlations with seven sets of experimental data, including flow pattern and tube inclination effects. Heat Transfer Eng 20(1):15–40

Kim D, Ghajar AJ, Dougherty RL (2000) Robust heat transfer correlation for turbulent gas-liquid flow in vertical pipes. J Thermophys Heat Transf 14:574–578

Kline SJ, McClintock FA (1953) Describing uncertainties in single sample experiments. Mech Eng 1:3–8

Lips S, Meyer JP (2012) Experimental study of convective condensation in an inclined smooth tube. Part II: inclination effect on pressure drop and void fraction. Int J Heat Mass Transf 55:405–412

Liu W, Tamai H, Takase K (2013) Pressure drop and void fraction in steam–water two-phase flow at high pressure. J Heat Transf 135:1–13

Lockhart RW, Martinelli RC (1949) Proposed correlation of data for isothermal two-phase, two component flow in pipes. Chem Eng Prog 45:39–48

Mandhane JM, Gregory GA, Aziz K (1974) A flow pattern map for gas-liquid flow in horizontal pipes. Int J Multiphase Flow 1:537–553

Marchaterre JF (1956) The effect of pressure on boiling density in multiple rectangular channels, Report ANL-5522, Argonne National Labs

Marchaterre JF, Petrick M, Lottes PA, Weatherland RJ, Flinn WS (1960) Natural and forced circulation boiling studies, Report ANL-5735, Argonne National Labs

Mishima K, Ishii M (1980) Theoretical prediction of onset of horizontal slug flow. J Fluids Eng 102:441–445

Mishima K, Ishii M (1984) Flow regime transition criteria for upward two-phase flow in vertical tubes. Int J Heat Mass Transf 27:723–737

Mukherjee H (1979) An experimental study of inclined two-phase flow, Ph.D. Thesis, University of Tulsa

Muller-Steinhagen H, Heck K (1986) A simple friction pressure drop correlation for two-phase flow in pipes. Chem Eng Process 20:297–308

Nguyen VT (1975) Two-phase gas-liquid cocurrent flow: an investigation of holdup, pressure drop and flow patterns in a pipe at various inclinations, Ph.D. Thesis, University of Auckland, New Zealand

Nicklin DJ, Wilkes JO, Davidson JF (1962) Two-phase flow in vertical tubes. Trans Inst Chem Eng 40:61–68

Oshinowo O (1971) Two-phase flow in a vertical tube coil, Ph.D. Thesis, University of Toronto, Canada

Owen DG (1986) An experimental and therotical analysis of equlibrium annular flow, Ph.D. Thesis, University of Birmingham, UK

Quiben JM, Thome JR (2007) Flow pattern based two-phase frictional pressure drop model for horizontal tubes. Part I: diabatic and adiabatic experimental study. Int J Heat Fluid Flow 28:1049–1059

Rezkallah KS, Sims GE (1987) An examination of correlations of mean heat transfer coefficients in two-phase and two-component flow in vertical tubes. AIChE Symp Series 83:109–114

Rouhani SZ, Axelsson E (1970) Calculation of void volume fraction in the subcooled and quality boiling regions. Int J Heat Mass Transf 13:383–393

Sacks PS (1975) Measured characteristics of adiabatic and condensing single component two-phase flow of refrigerant in a 0.377 inch diameter horizontal tube, ASME Winter Annual Meeting, Houston, 75WA/HT-24, pp 1–12

Schlegel J, Hibiki T, Ishii M (2010) Development of a comprehensive set of drift flux constitutive models for pipes of various hydraulic diameters. Prog Nucl Energy 52:666–677

Seider EN, Tate GE (1936) Heat transfer and pressure drop of liquids in tubes. Ind Eng Chem 28:1429–1435

Shannak BA (2008) Frictional pressure drop of gas liquid two-phase flow in pipes. Nucl Eng Des 238:3277–3284

Shedd TA (2010) Void fraction and pressure drop measurements for refrigerant R410a flows in small diameter tubes, Technical Report AHRTI 20110-01

Shoukri M, Hassan I, Gerges I (2003) Two-phase bubbly flow structure in large diameter vertical pipe. Can J Chem Eng 81:205–211

Spedding PL, Che JJJ, Nguyen NT (1982) Pressure drop in two-phase gas-liquid flow in inclined pipes. Int J Multiphase Flow 8:407–431

Spedding PL, Woods GS, Raghunathan SR, Watterson JK (2000) Flow pattern, holdup and pressure drop in vertical and near vertical two and three phase upflow. Chem Eng Res Des 78 (Part A):404–418

Taitel Y, Dukler AE (1976) A model for predicting flow regime transitions in horizontal and near horizontal gas-liquid flow. AICHE J 22:47–55

Taitel Y, Barnea D, Dukler AE (1980) Modeling flow pattern transitions for steady upward gas-liquid flow in vertical tubes. AICHE J 26:345–354

Tang CC (2011) A study of heat transfer in non-boiling two-phase gas liquid flow in pipes for horizontal, slightly inclined and vertical orientations, Ph.D. Thesis, Oklahoma State University

Tang CC, Ghajar AJ (2011) A mechanistic heat transfer correlation for non-boiling two-phase flow in horizontal, inclined and vertical pipes, Proceedings of AJTEC 2011 ASME/JSME 8th Thermal Engineering Joint Conference, Paper No. AJTEC2011-44114

Terekhov VI, Pakhomov MA (2008) The effect of bubbles on the structure of flow and the friction in downward turbulent gas-liquid flow. High Temp 46:854–860

Usui K, Sato K (1989) Vertically downward two-phase flow: (I) void distribution and average void fraction. J Nucl Sci Technol 26(7):670–680

Wallis GB (1969) One-dimensional two-phase flow. McGraw-Hill, New York

Weisman J, Kang SY (1981) Flow pattern transitions in vertical and upward inclined lines. Int J Multiphase Flow 7:271–291

Woldesemayat MA, Ghajar AJ (2007) Comparison of void fraction correlations for different flow patterns in horizontal and upward inclined pipes. Int J Multiphase Flow 33:347–370

Wongs-ngam J, Nualboonreung T, Wongwises S (2004) Performance of smooth and micro-finned tubes in high mass flux region of R-134a during condensation. Int J Heat Mass Transf 40:425–435

Wongwises S, Pipathttakul M (2006) Flow pattern, pressure drop and void fraction of gas–liquid two-phase flow in an inclined narrow annular channel. Exp Thermal Fluid Sci 30:345–354

Xiong R, Chung JN (2006) Size effect on adiabatic gas-liquid two-phase flow map and void fraction in micro-channels, Proceedings of Int. Mech. Eng. Congress and Exposition, Chicago

Xu Y, Fang X (2012) A new correlation of two-phase frictional pressure drop for evaporating two-phase flow in pipes. Int J Refrig 35:2039–2050

Yijun J, Rezkallah K (1993) A study on void fraction in vertical co-current upward and downward two-phase gas-liquid flow – I: experimental results. Chem Eng Commun 126:221–243

Zhang W, Hibiki T, Mishima K (2010) Correlations of two-phase frictional pressure drop and void fraction in mini-channel. Int J Heat Mass Transf 53:433–465

Zuber N, Findlay J (1965) Average volume concentration in two-phase systems. ASME J Heat Transf 87:453–468

Index